Applied Design Research: The Societal Impact

Applied design researchers are at the forefront of initiating and executing projects to catalyze societal change. But how exactly do they accomplish this? What are the various ways in which design research can contribute to societal transitions? How do these researchers bridge different system levels through their projects? Furthermore, how do they formulate research questions and deliver outputs that leave a lasting impact on society, and finally, how do design researchers navigate the delicate balance of incorporating different worldviews into their projects, while also considering their own perspectives? This book provides insightful answers to these questions, drawing on the expertise and experiences of established design researchers in the field.

Applied Design Research: The Societal Impact discusses the transformative power of design research in driving societal change. This publication offers valuable insights into addressing pressing global issues such as the circular economy, healthcare challenges, and sustainability. The book explores the multifaceted role of design in influencing various levels of systems, including materials, products, services, and socio-technical ecosystems. Examining the underlying theories of change illuminates how design interventions can spark significant ripple effects within broader societal change processes. Through a collection of practical examples and case studies, this book showcases how applied design research can yield tangible and enduring societal impact.

Edited by Koen van Turnhout [0000-0002-7360-6156], Peter Joore [0000-0002-4585-008X], Remko van der Lugt [0000-0002-5920-4126], Troy Nachtigall [0000-0002-2136-1214] and Liliya Terzieva [0000-0003-4834-6951].

I0041816

Edited by
Koen van Turnhout, Peter Joore,
Remko van der Lugt,
Troy Nachtigall, Liliya Terzieva

CRC Press
Taylor & Francis Group

Colophon

First Edition Published 2026
By CRC Press
2385 Executive Center Drive, Suite 320, Boca Raton, FL, 33431, USA

and by CRC Press
4 Park Square, Milton Park, Abingdon, Oxon, OX14 4RN

CRC Press is an imprint of Taylor and Francis Group, LCC

This project was co-funded by Taskforce Applied Research SIA, part of the Netherlands Organisation for Scientific Research (NWO). Open access of this publication was financed by NWO. NWO Grant ID: https://doi.org/10.61686/KUARK01421.

Taskforce for Applied Sciences SIA / Dutch Research Council NWO.

ISBN: 978-1-041-00424-0 (hbk)
ISBN: 978-1-041-00020-4 (pbk)
ISBN: 978-1-003-60976-6 (ebk)
DOI: 10.1201/9781003609766

Publishers note: This book has been prepared from camera-ready copy provided by the authors.

Names
Van Turnhout, Koen, editor. | Joore, Peter, editor | Van der Lugt, Remko, editor | Nachtigall, Troy, editor | Terzieva, Liliya, editor

Title
Applied Design Research: The Societal Impact / edited by Koen van Turnhout, Peter Joore, Remko van der Lugt, Troy Nachtigall, Liliya Terzieva

Description
First edition. | Boca Raton, FL: CRC Press, 2025. | Includes bibliographical references.

Funding
This project was co-funded by Taskforce Applied Research SIA, part of the Netherlands Organisation for Scientific Research (NWO). Open access of this publication was financed by NWO.
NWO Grant ID: https://doi.org/10.61686/KUARK01421.

Contributors
Koen van Turnhout, Peter Joore, Remko van der Lugt, Troy Nachtigall, Liliya Terzieva, Daan Andriessen, Danielle Arets, Bas de Boer, Aranka Dijkstra, Agnes Evangelista, Poorvi Garag, Christa van Gessel, Jetse Goris, Kees Greven, Risk Hazekamp, Jan-Wessel Hovingh, Tomasz Jaskiewicz, Derek Kuipers, Elise van der Laan, Mailin Lemke, Marjolein Mesman, Catelijne van Middelkoop, Lenny van Onselen, Mieke Oostra, Anja Overdiek, Kim Poldner, Margo Rooijackers, Angelique Ruiter, Shakila Shayan, Wina Smeenk, Iskander Smit, Guido Stompff, Gijs Terlouw, Jet van der Touw, Lars Veldmeijer, Sophie Vermaning, Nick Verouden, Rens van der Vorst, Inge Vos, Jeroen de Vos, Roelof de Vries, Judith Weda, Bart Wernaart, Tamara Witschge, Marieke Zielhuis, Banoyi Zuma

Design and Layout
Studio RATATA.nl

Illustrations and Cover Image
Kalle Wolters

**network
applied
design
research**

NDR2501_CRC_EN16

Content

Part 1: Theory of Change

Part 2: Connecting System Levels

Part 3: Balancing Worldviews

Part 4: Beyond Solutionism, Design without End

Preface

Design has the power to create change—but how do we ensure that change truly matters? That question lies at the heart of this publication and the Network Applied Design Research (NADR). As Chair of NADR, I am very enthusiastic to introduce this third international publication of our network.

Since its founding in 2015, NADR has brought together design researchers and practitioners from diverse domains such as product design, healthcare, architecture, food, retail, multimedia, and fashion. What connects them is not only their use of design as a research approach, but their shared ambition: to contribute to positive, systemic shifts in society.

This publication is the result of our annual knowledge cycle, in which partners of NADR exchange insights and experiences around a central theme. Following earlier editions that provided a general overview of the field of applied design research (Joore et al., 2022) and explored living labs and other experimental learning and innovation environments (Joore et al., 2024), this third publication tackles perhaps the most fundamental question: how can applied design research contribute to real societal impact? How can design practice support systemic change and transitions towards, for example, a circular economy, smarter healthcare, or sustainable fashion?

Across 21 chapters, 46 authors—representing a broad spectrum of disciplines and perspectives—explore this question from multiple angles. Their contributions make it clear that there are no simple answers. They address challenges such as linking developments across system levels, the limited makeability of society, and navigating the diverse worldviews of those involved. At the same time, they highlight the potential of design as a catalyst for meaningful transformation. What emerges is not a blueprint for impact, but a collective reflection on what it means to try—and keep trying—to design for it.

I invite you to explore these insights, learn from the experiences shared, and apply them in your own practice—so we can continue creating positive societal change, together.

Peter Joore

Chair, Network Applied
Design Research

• Joore, P., Stompff, G., & van den Eijnde, J. (Eds.). (2022). *Applied design research: A mosaic of 22 examples, experiences and interpretations focussing on bridging the gap between practice and academics.* CRC Press.
• Joore, P., Overdiek, A., Smeenk, W., & van Turnhout, K. (Eds.). (2024). *Applied design research in living labs and other experimental learning and innovation environments. CRC Press.*

Koen van Turnhout,
Peter Joore, Remko van der Lugt,
Troy Nachtigall, Liliya Terzieva

1. Sweet Spots and Orchestrations: Examining the Societal Impact of Applied Design Research

DOI: 10.1201/9781003609766

Introduction

As applied design researchers we are sensemakers: we are digesting society's complexity and set out to improve it in meaningful ways. With our practices, we aspire to be catalysts for societal transformation. We apply our imaginative abilities to come up with alternatives for the status quo and we materialise ideas into tangible outcomes, not so much to solve things, but to set things in motion.

The central question of this book is focussed on the "how" of that practice. How can applied design research projects have an impact on society, what are the challenges that we face in achieving this impact, and how do we overcome those challenges? How do we ask the right questions, how do we transcend boundaries in society, or within our own projects? How can design function as means and not as an end? In this volume, over thirty authors working in diverse fields such as retail, human-computer interaction, architecture, healthcare, education, behaviour change, communication or bio-design reflect on such questions. Together they show the multifaceted nature of applied design projects and their societal contributions. There are many pathways to change, transform and enable transitions and many ways to navigate those. We hope to show the richness of our collective thinking about impact, and we invite applied design researchers to pick up this book to reexamine their own views and assumptions, and to seize the opportunity to learn from each other. This reflective journey not only enhances the rigor of applied design research but also empowers practitioners to make more meaningful and sustainable contributions to society.

This book is divided into four themes. First we feature chapters about the theory of change of applied design researchers. Contributors offer viewpoints on the pathways of societal impact, and how they apply this to their own projects. Next we address the complexity of society. Contributors discuss how they manage to connect system levels and reflect on how changes at one systemic level can affect others. From that we move to complexities within projects. In this section, authors discuss how they manage to balance worldviews within and beyond their projects. Applied

10

design research almost invariably involves different epistemic cultures. Participants differ in what they view as good ways to obtain knowledge within a project and what knowledge can be trusted. Navigating these differences is a necessary step towards impact. Finally we discuss the ways in which applied design research can have an impact beyond solving practical problems. Authors reflect on their work within the theme beyond solutionism, design with no end. After reading this work, one thing will become clear: applied design researchers aren't just sensemakers; we create to facilitate sensemaking around us and beyond.

Part 1: Theory of Change

This book invites researchers to challenge their assumptions about the way in which their projects contribute to society. In the first part of the book we feature work that tackles this problem head-on. We provide a different perspective to researchers to think about the pathway that connects their project activities to the impact in society that it intends to make. This is also referred to as impact and Theory of Change. Many researchers tend to have little or no idea of how their activities are really going to lead to societal change, and therefore their efforts are difficult to evaluate. It is therefore necessary to highlight and articulate the assumptions that underlie the researchers' actions and to look for shorter-term changes that could point at whether these assumptions hold ground.

Even though the concept of Theory of Change originates in the field of project evaluations, ideas about change are important in all stages of research: they govern how projects are conceived, how they are executed, and how we look at what happens 'after' a project. Constructing a theory of change can help researchers to uncover their assumptions of these different machineries of change. At the same time, it is important to be aware that it is what it is: a theory. The complexity of the societal issues that applied design researchers address make things unfold quite differently in reality. So a 'theory of change' helps to uncover presuppositions, but also requires adjustment over time as insight grows. The purpose of this section is to ignite thinking about these assumptions, and become more critical to them as applied design researchers.

In the first chapter, Framing the Impact of Applied Design Research as Enhancing Practice Communities' Solution Repertoire. Is it Helpful?, Koen van Turnhout, Wina Smeenk, Marieke Zielhuis, & Daan Andriessen give an overview of the 'impact literature' in science. They discuss various critiques of the dominant way of looking at impact and suggest that Applied Design Research is in need of its own impact models. They propose an alternative based on the notion of solution repertoires. In their view, evaluating applied design research projects deals with examining how the solution repertoires of the different target practitioner groups change through the project. They explain the approach by means of a dementia simulator case example.

The first step in conceiving projects with impact is probably addressing the right research questions or challenges. This question is tackled in the second chapter, The Impact of Technology Maturity on Determining Urgent Research Questions. Roelof de Vries, Judith Weda, Shakila Shayan and Koen van Turnhout propose a model for determining urgent research questions, based on Koert van Mensvoort's Pyramid of Technology. They describe how (technology-based) design research deals with transitions in levels of naturalization: the level to which a technology moves from being merely envisioned to having become part of human nature. Thus, considering the trajectory of technology can provide a valuable component in developing a theory of change.

Obviously much impact is also made during research projects, often by seducing stakeholders into novel ways of thinking. In Co-designing Potential Boundary Objects to Open Patient Perspectives and Redressing Power Imbalances in Mental Healthcare Design Studies, Gijs Terlouw and Lars Veldmeijer describe how artefacts can be designed with the potentiality of functioning as boundary objects. They stress that the designed healthcare interventions intended to allow for meaningful interactions between caregiver and patient/client without the need to fully understand each other or to reach consensus.

This notion of boundary crossing as an important precondition for applied design research projects to have an impact is taken even a step further by Derek Kuipers, Jan-Wessel Hovingh and Jetse Goris in their chapter, No I Don't Know Either - New Rituals for Collective Learning: The Design Process as a Temporary Boundary World. These authors regard co-design processes as dwelling within a 'temporary boundary world'. They see design as a collective process of learning and adaptation, rather than a means of solution production. They demonstrate their approach in a case with a primary school in a rural area with a declining population.

The last contribution in this section bridges the impact within project to the impact after projects by focussing on how co-design practices can bleed into professional practices that are not engaged in design such as health care. In Impact by Toolkits as Generative Knowledge Carriers in Healthcare, Remko van der Lugt, Lenny van Onselen and Christa van Gessel reflect on their experiences with creating toolkits for healthcare professionals as a means to transfer research insights to practice. Such toolkits are a way to solidify intermediate knowledge. They see toolkits as generative knowledge carriers, which means that they are not only intended to transfer the knowledge, but also to spur ideation and making new connections, to enable health care practitioners to integrate such insights in designerly ways.

Part 2. Connecting System Levels

Design research today is no longer confined to discrete objects or narrow user interactions; it actively shapes and responds to interconnected systems of people, environments, and technologies. In the second part of this book we focus on the ways in which applied design researchers try to tackle this societal complexity. The contributions in this part illustrate how systemic thinking permeates all aspects of design. Each of the chapters manifests how designers apply systemic knowledge to create societal impact, connecting multiple system levels. From micro-level interventions to macro-level systemic transitions, design researchers are tasked with addressing challenges that span personal, local, and societal scales. Such multi-layered engagement is crucial to navigate large-scale transitions like climate change, urban development, and health innovation. A recurring insight is that effective design interventions need to engage at different system levels simultaneously, enabling cross-level interactions between local interventions and institutional frameworks. By acting at multiple scales, design not only contributes to the development of localised solutions, but may also help develop pathways to systemic shifts, scaling and accelerating.

Jet van der Touw and Anja Overdiek start their chapter, Leverage Points Revisited: How Social Designers (Can) Apply Systemic Knowledge to Make Impact, by introducing a working definition of 'system' as a collection of interconnected elements working towards a common purpose. In their research, they draw upon insights from broader systems theory, particularly the concept of leverage points, which they define as places in complex systems where a small shift may lead to fundamental changes in the system as a whole. Their approach identifies four levels at which designers can intervene, which they tested in a case of policy advice for the Dutch Ministry of Internal Affairs. When testing their model, they conclude that simply asking the right questions helps designers to identify the areas of the system where they can have the greatest impact.

In the chapter by Elise van der Laan, Agnes Evangelista, Kim Poldner and Marjolein Mesman, Harnessing Retail Staff as Sustainable Fashion Influencers: A Pathway to Systemic Societal Impact through Applied Design Research, the ambition of the design researchers is to make the entire fashion industry more sustainable. Naturally, this system is far too large to tackle within a single design project. Therefore, the designers focus on the level of Multi-Brand Fashion Retailers, where they explore how these can harness retail staff as influencers to promote engagement with sustainable fashion. During the project, they zoom in until they reach the smallest micro level, where they design small clothing labels with QR codes that allow customers to access information about their garments. Their research revealed that these micro interventions work best in combination with training and organisational support on higher levels of organisation.

In her chapter, Unravelling Lifeworlds: Farmers and Consumers of Black Pepper, Poorvi Garag explores the international pepper trade system, examining not only the current system but also its historical development and its relationship with colonial history. Garag conducts interviews with two distinct groups: Western consumers who purchase a jar of pepper in the supermarket, and local producers of black pepper in India. By doing so, she connects production and sales on a global scale, demonstrating how a simple peppercorn is, in fact, part of a highly interconnected global system. To raise awareness among consumers about the value of the Indian origins of black pepper, Garag devised an intervention based on speculative fabulation, using storytelling as a method to explore alternative futures and imagine a radically different world. In her imagined world, black pepper remains a currency, a 'black pepper wallet' stimulates discussion at various exhibitions. A simple design fiction like this may create a shift in perspective, fostering a deeper understanding of the origins of black pepper, international trade, and colonial history.

The final chapter of this section, My Life as a Living Lab Coordinator, by Peter Joore and Aranka Dijkstra, also adopts a narrative approach, albeit for a somewhat less speculative project. This chapter describes how the AMS Institute in Amsterdam employs

Urban Living Labs to implement complex changes in the urban environment. It is written from the perspective of a Living Lab Coordinator, whose role is to initiate the intended innovation in collaboration with all relevant stakeholders. The chapter focuses on a local neighbourhood that already generates a significant amount of sustainable energy but aims to develop storage solutions to balance the difference between energy production and consumption. The narrative clearly highlights the challenges the Living Lab Coordinator encounters throughout this process. The underlying message of the chapter is that the Living Lab Way of Working, developed by AMS Institute, serves as a valuable tool to help steer processes in an Urban Living Lab in the right direction.

Part 3. Balancing Different Worldviews

In the third section of this book, we turn our attention to the people involved in applied design research projects, and in particular to how they see the world. If applied design research is to have a profound impact on societal transitions, a delicate orchestration of diverse worldviews, priorities, and methodological approaches is a necessary precondition. As design researchers tackle complex challenges ranging from behaviour change to urban development, they need to navigate a landscape where different disciplines, cultures, and stakeholders intersect.

Balancing different worldviews in applied design research represents a critical challenge and opportunity in generating meaningful societal impact. The complexity of this balance manifests in several key dimensions. First, researchers need to reconcile academic rigour with practical applicability, ensuring their work meets scholarly standards while remaining relevant and accessible to communities and practitioners. Second, they need to bridge the gap between systematic design approaches and the often messy, unpredictable nature of social change. Third, they have to balance the drive for innovation with respect for existing cultural practices and local knowledge systems (Björgvinsson et al., 2012).

Applied design researchers need to mediate between different stakeholder priorities, from community needs to institutional requirements, while maintaining the integrity of the design process (Bannon and Ehn, 2013). They have to consider how different worldviews influence not just the design outcomes, but also the very process of collaboration and co-creation. This includes understanding how power dynamics, cultural differences, and varying definitions of 'impact' shape project trajectories and outcomes. It requires continuous reflection, adaptation, and a willingness to challenge one's own assumptions and methods. The contributions in this section explore how researchers go about this in practice.

In their chapter, Research Culture Clash in Behaviour Change Design: Three Dilemmas for Behaviour Change, Bas de Boer, Roelof de Vries and Mailin Lemke, for example, examine how differing research cultures, rather than academic or funding structures, create fundamental dilemmas in behaviour change design. Three differences in epistemic culture are the perception of problem complexity, process approaches, and impact assessment methods. By making these cultural differences explicit, the chapter provides a framework for researchers to better understand and navigate interdisciplinary collaboration in behavior change design projects.

Next, Cathelijne van Middelkoop explores how artistic research practices can enhance applied design research. Her chapter, The Transformative Powers of Artistic Research within Design Education, explores this challenge through a case study of the Practorate Meaningful Creativity at Sint Lucas, demonstrating how artistic approaches foster critical reflection and engage with broader philosophical questions beyond traditional problem-solving. Through the examination of visual markers and embodied knowledge, the study reveals both the potential and challenges of integrating artistic research into design processes, particularly in vocational education, while highlighting the need for further methodological development to validate and share tacit knowledge.

17

Co-creation is often proposed as a way to include very different perspectives within a design project, but this comes with its own challenges. This is explored by Nick Verouden, Sophie Vermaning and Tamara Witschge in their chapter, Navigating Inclusion and Ethical Challenges in Co-Creation. They reflect on the challenges of co-creation in the digital transformation of urban spaces, highlighting the gap between its democratic ideals and practical constraints. Through an ethnographic study of the CommuniCity project, it underscores the need for flexible, reflexive approaches that prioritize meaningful community engagement, long-term collaboration, and context-sensitive adaptation to enhance the transformative potential of co-creation.

This thread is picked up by Banoyi Zuma, Liliya Terzieva, Margo Rooijackers and Inge Vos. In Matching and Mismatching Worldviews Entangled in Art – Applied Design Research in "Hip-Hop Dance" Towards Impact, they investigate how personal values and ideologies influence design research initiatives through examining the Inclusive Dancing project, which uses hip-hop dance as a medium to study inclusion between children with and without special needs. Through analyzing the pilot phase data from hip-hop dance classes, the research explores reflexivity awareness in design research while assessing cognitive and emotional well-being through an inclusion impact framework.

Finally, Kees Greven en Daan Andriessen investigate how collaboration in interdisciplinary applied design research projects can be fostered by paying attention to differing worldviews. In The Use of a Research Paradigm Workshop for Embracing Different Worldviews in Practice-Based Research, they illustrate making the differing values, norms, and assumptions of stakeholders, explicit in a workshop a shared language for discussing research design and methodologies, can be established. Ultimately, this promotes mutual understanding and respect in projects that bridge disciplines such as healthcare and design.

_____ Koen van Turnhout, Peter Joore, Remko van der Lugt, Troy Nachtigall, Liliya Terzieva

Part 4. Beyond Solutionism, Design without End

Design research has evolved considerably over the past few decades, moving from a seemingly straightforward process of identifying problems and delivering solutions, toward a more exploratory, emergent, and reflective practice. As such, the series of chapters in this part explore design as an unfolding process, anchored in continuous reflection, making, and discovery.

This part emphasizes the ways in which design is a continuous, emergent act—a journey "without end," as the design process rarely concludes with a definitive answer. Instead, design involves the willingness to embrace emergence and to engage in creative entanglements with materiality. This perspective foregrounds an ongoing, open-ended approach that seeks interpretive frameworks for understanding and reimagining the world, rather than final solutions.

Embedded in the core of design is an assumption of make-ability— the notion that no matter how convoluted the situation, creative possibilities always exist. However, the scale and complexity of societal transitions complicate this assumption, since problems continuously shift and resist neat closure. Nevertheless design—as a craft and as a research practice—has great potential to address such societal challenges. It is exactly the qualities that make design distinctive—craftsmanship, playfulness, and an openness to the unexpected—that are needed for impact in an ever-changing highly complex world.

This section starts with Design in the Real World: Designing beyond the Ends of Design by Guido Stompff, Tomasz Jaskiewicz, Troy Nachtigall, Iskander Smit. This chapter expands the Beyond Solutionism discourse by emphasizing design as a continuous, collective learning process rather than a means of problem-solving. Through the concept of Fifth Order Design, it reframes prototypes as dynamic boundary objects that provoke engagement and emergent meanings. The chapter challenges linear design methodologies, advocating for open-ended, evolving interventions that foster co-creation and systemic transformation.

19

Zooming out a bit, An Acupuncture Approach for Societal Change: Reflections on the Make-ability of Society, by Mieke Oostra & Koen van Turnhout, introduces research on acupuncture as a method for navigating complex societal challenges through small, strategic design interventions. Drawing on urban acupuncture, the authors explore how targeted projects—such as retrofitting high-rise buildings—can drive systemic change despite resource constraints and shifting contexts. Their work challenges the assumption of make-ability, advocating instead for strategic navigation: an adaptive, iterative approach that continuously responds to emerging opportunities and uncertainties in design.

Social Change is addressed next. Design Is Not the Answer! Reflective Practices and the Social Field: A Case Study in Amsterdam Gaasperdam challenges conventional design-led approaches to societal transitions. Angelique Ruiter argues that meaningful change in the energy transition requires strengthening the social field—the quality of relationships, trust, and shared understanding within a community. Through a two-year study in Gaasperdam, she demonstrates how reflective practices and deep engagement foster community-driven action. This chapter highlights the need for designers to move beyond solutionism, embracing participatory approaches that centre lived experience and emergent social dynamics.

Crafting Conflict: Designing Meaningful Interactions in a Smooth Society argues for the necessity of designing friction in the context of smart city developments. Danielle Arets, Bart Wernaart, Jeroen de Vos and Rens van der Vorst critique the solutionist and technocratic tendencies of smart cities, advocating instead for adversarial design as a method to surface conflicting values and ethical dilemmas. Through the Moral Design Game, they explore how structured dissent can foster democratic engagement, making hidden tensions in smart technologies more tangible. This chapter highlights the role of design in enabling critical discourse, rather than smoothing over conflict.

We end with "You Press the Button, We Do the Rest." – How to Cede Agency to the More-than-Human? explores the shift from make-ability to response-ability in bio-art and bio-design. Risk Hazekamp et al. challenge human-centric notions of authorship by engaging with *Synechocystis* sp. PCC 6803, a cyanobacterium, to co-create living photographic processes. Through autoethnographic reflection and experimental practice, their chapter investigates how unlearning, uncertainty, and multispecies collaboration can reframe artistic research as an emergent and relational practice, decentring human agency in favour of more-than-human co-authorship.

Part 1:
Theory of Change

"Ideas about change are important in all stages of research: they govern how projects are conceived, how they are executed and how we look at what happens 'after' a project."

Koen van Turnhout, Wina Smeenk,
Marieke Zielhuis, Daan Andriessen

2. Framing the Impact of Applied Design Research as Enhancing Practice Communities' Solution Repertoire. Is It Helpful?

DOI: 10.1201/9781003609766

Introduction

Applied design researchers engage in complex projects that have diverse forms of impact on multiple stakeholders. One could think, for example, of the dementia simulator by, amongst others, Wina Smeenk (2022). The project, aimed at improving the public's understanding of the experiences of—and empathy with—people with dementia, yielded results for different groups at different levels and timelines. The resulting simulator helped informal and formal caregivers, societal stakeholders such as police and supermarkets, and the general public understand dementia. Moreover, the work on the simulator involved improving empathic design methodology (i.e. Smeenk et al. 2018, Smeenk 2019), which inspired and informed design researchers designing for such sensitive situations worldwide. Moreover, all kinds of stakeholders around dementia, such as care organisations, government bodies, and the health(care) industry, changed how they thought about the dementia problem because of the existence of and their experience within the simulator, and discussions later on (Spek, Sleeswijk & Smeenk, 2024). If so, many stakeholders are involved and their 'learnings' are so diverse and often implicit, how do we establish the societal impact of such a project?

In this chapter, we aim to unravel and articulate the various ways Applied Design Research (ADR) projects can make an impact. While numerous approaches for evaluating the impact of (applied) research exist, it is easy to become overwhelmed by the multitude of philosophical underpinnings, approaches, and criteria for impact or research efforts that have been proposed. Current frameworks also pay little attention to the unique contribution of ADR: exploring alternatives for the status quo in society. This paper suggests that the notion of 'Solution Repertoire' (van Turnhout & Smits, 2021) can provide a viable and practical alternative to some of the other approaches for impact evaluation.

The chapter is structured as follows. First, we examine an existing NWO impact model based on the theory of change and the academic discussion about this model. Then, we introduce the notion of *Solution Repertoire* and illustrate it with an example. Next, we analyse how an evaluation approach based on Solution Repertoire can form an alternative to other impact frameworks.

Requirements for Impact Evaluation Approaches

The impact of research efforts on society is fiercely debated. Many academics desire to contribute to society, and many governmental funding programs are set up to stimulate research efforts that strengthen society. As such, there is a need for models that help scientists explicate the impact of their research efforts. In order to understand what an impact evaluation approach for Applied Design Research needs to accomplish, it is important to get a sense of the relevant elements of impact that models and approaches identify. This leads into requirements that can be used to evaluate our own approach. In this section, we extract such requirements (for impact models) by discussing both the most dominant impact model in the Netherlands and its critiques in the literature.

The most well-known impact model in the Netherlands is most likely the Impact Pathway Model (IPM) required by the Dutch funding agency NWO for most projects (NWO, n.d.); see Figure 2.1. The model is based on the theory of change approach (i.e. Weiss, 1997), which is common in impact evaluations worldwide. The idea is to encourage researchers to build a chain from the societal challenges they would like to address via underlying causes to research questions and then back from research output via research outcomes to societal impact. This depicts research impact as a layered process in which the impact gets relayed to a bigger group in each step: the consortium has an impact on the stakeholders, who, in turn, have an impact on society.

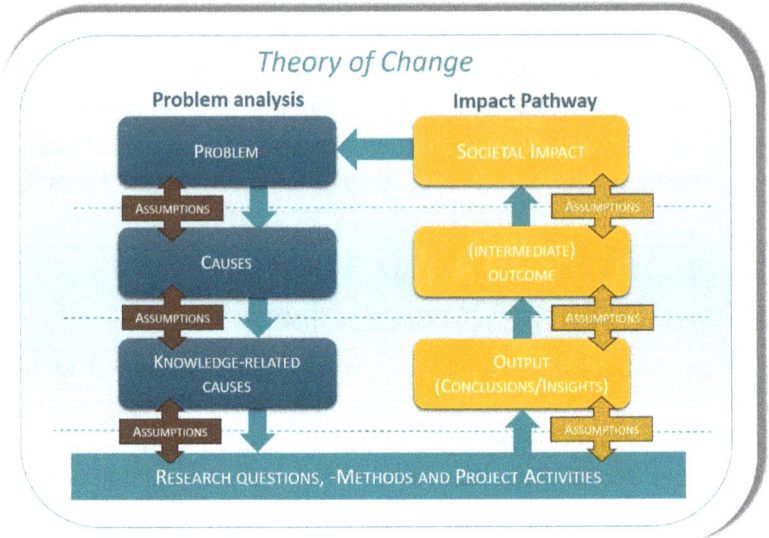

Figure 2.1. The NWO Impact Pathway Model, based on the theory of change (NWO, n.d.).

However, this model has been criticized in the literature. Van Vliet (2022) summarizes the criticism into three points.

- The Impact Pathway Model suggests that impact is achieved after the research has been finished rather than in the process of co-creation during the project. It suggests that research and dissemination are different 'phases', whereas many practice-oriented projects already have an impact when the research is still ongoing.
- The model suggests a single stream of knowledge transformation rather than a set of parallel processes with different social dynamics. The research process may have an immediate impact; some elements of the research might have a quick impact while the impact of other elements is delayed.
- The model suggests stakeholders agree on the problem and utilize the result in the same way, while in reality, there are many value conflicts within and around any research project.

Such critiques can be easily dismissed by pointing to the fact that the impact pathway model is 'just' a model and thus naturally abstracts away from the complexities of 'in vivo' research activities. However, this raises the question of to what extent the elements captured in the model are the most essential for understanding research impact. Other literature suggests several requirement categories not currently in the model: target community, timelines, pathways, notion of knowledge, and impact types.

The first is the *target community*. Research projects typically serve multiple audiences. The NWO impact model summarises these under 'stakeholders', which is not very specific. Greven and Andriessen (2019) are more precise by suggesting that *practice*, *education*, and *science* are essential audiences to consider in practice-based projects. Practice, in itself, is a fairly broad category, though. This could include professional groups, such as professional designers, but also domain professionals in, for example, healthcare or education and policymakers (Zielhuis, 2023). These groups, in turn, sustain their knowledge in practitioner networks. As such, Van Turnhout and Smits (2022) urge us to identify opportunities for knowledge buildup in the different practitioner networks a project touches.

Second, we need to consider *timelines*. It is often alluded that scientific impact is slow (Chakrabarti & Lindeman, 2016). There are multiple processes involved that influence this speed. First, it can be attributed to the speed of scientific progress—for example, because knowledge buildup in the scientific community is slow. Second, it can be attributed to the speed of the dissemination, for example, through students who bring knowledge into practice only after they graduate. Third, it can be attributed to adoption speeds in organisations: startups, for example, are quicker to adopt new ideas than big organisations (Arlitt et al., 2016). One could argue that Applied Design Research escapes from some of these hurdles because it involves direct collaboration of stakeholders, and as such, these projects directly impact practice. However,

29

ADR projects need to have an impact beyond the partners directly involved in the project. We also need to consider the timelines of knowledge buildup in the knowledge networks of these other stakeholders.

The third issue is *pathways*. How, exactly, does knowledge that is created in research projects end up in practice? Part of the reason the impact puzzle is so complex is that there are many pathways. The NWO impact model abstracts away from these pathways by summarizing them into 'productive interactions', which can be specified for each block of the model in Figure 2.1. The idea is that research exercises (societal) impact through co-creation between researchers and stakeholders outside academia (Spaapen & van Drogen, 2011). Productive interactions can further be distinguished between financial interactions (i.e. funding research), direct interactions (meeting points between people), and mediated interactions (interactions through knowledge products such as books or tools).

This notion of productive interactions reduces diverse pathways for research impact, such as collaboration within ADR projects, students and researchers that enter practice, training courses, and practitioner networks into easily understood categories. However, since impact happens in parallel processes with different timelines, it seems necessary to distinguish between somewhat fleeting *modes of impact* – such as direct collaboration – and how knowledge gets sustained, such as in practitioner networks, through knowledge products, and educational programs. Moreover, the pathways generally identified are 'from research to practice', whereas the reverse route is also essential to breed relevant research (Smeenk et al., 2024).

A fourth issue concerns the underlying *notion of knowledge* in an impact framework. Vliet (2022) identifies four *epistemic viewpoints* on knowledge, each suggesting distinct approaches to evaluating impact. First, the positivist viewpoint sees knowledge as objective facts uncovered by research and disseminated across communities, with impact traced through the spread of facts. Second, the constructivist viewpoint views knowledge as constructed within specific historical and socio-cultural contexts, requiring impact to be measured by tracking local sensemaking. Third, the performative viewpoint suggests knowledge emerges from actor-network practices, with impact identified by changes in organisational practices. Fourth, the realist viewpoint defines knowledge by its fit to a specific context, with impact traced through translations into that context. Given the fundamental differences between these perspectives, it appears unwise to use an impact model that does not align with a specific epistemic viewpoint (Coombs & Meijer, 2021).

A fifth issue is whether *knowledge exchange* is the only way to reach impact. Greven and Andriessen (2019) suggest that there are other *impact modes*. They identify four ways in which practice-based projects have an impact: through the learning of individuals, through the creation of products, through the creation of knowledge, and system change. This is particularly relevant for applied design research projects, where the design of products and services and the achievement of system changes are more rules than exceptions. Table 2.1 summarises these impact aspects and the lessons we can draw for an impact assessment approach relevant to ADR projects.

Category	Suggested categories	Requirements
Target Community	Practitioners (i.e. designers, policymakers, domain professionals, etc.), education, science	Account for: -Practitioners are a diverse group (Zielhuis, 2023) -Knowledge is sustained in multiple networks (Van Turnhout & Smits, 2022).
Timelines	Speed of buildup, speed of dissemination, speed of adoption.	Knowledge build-up is different in different (practitioner) communities that projects contribute to (Van Turnhout & Smits, 2022), and these need to be addressed separately.
Pathways	Direct collaboration, methods and tools, training courses, educational programs, people, scientific and practitioner networks	Separate direct impact in the project from how knowledge is sustained in a practitioner community.
Notion of knowledge	Performative, positivist, realist, constructivist	The notion of knowledge behind the impact framework needs to be explicated (Coombs & Meijer, 2021)
Impact modes	Knowledge, learning, system change, products.	Account for multiple impact modes employed by ADR projects.

Table 2.1. Aspects of impact and lessons learned from the literature for constructing an ADR-specific impact evaluation approach.

The Notion of Solution Repertoire

This chapter examines whether the notion of *Solution Repertoire* can be a viable alternative for impact evaluation based on the theory of change. Van Turnhout and Smits (2021) introduced solution repertoire in the context of design education. They defined it as "the competence of appreciating and handling solutions in one's design discipline".

Van Turnhout and Smits build on work on design cognition that suggests designers have a solution-oriented reasoning style in which existing designs—called precedents—play a central role (Lawson, 1979; Pasman, 2003). Instead of a hierarchical model of design knowledge in which concrete designs are 'merely' an illustration of an underlying theory, van Turnhout and Smits suggest that concrete designs 'embody' design knowledge.

The idea is that designers have experience in and knowledge of concrete solutions to existing problems and understand those through multiple theoretical perspectives. In this view, abstract design knowledge such as design philosophies, heuristics, aesthetics, and predictive design theories are a glue between different design alternatives (Figure 2.2). A designer proposes a design alternative as a variant of an already known design solution through several theoretical lenses, each of which brings to mind a family of alternative designs the designer can adapt to enrich the design.

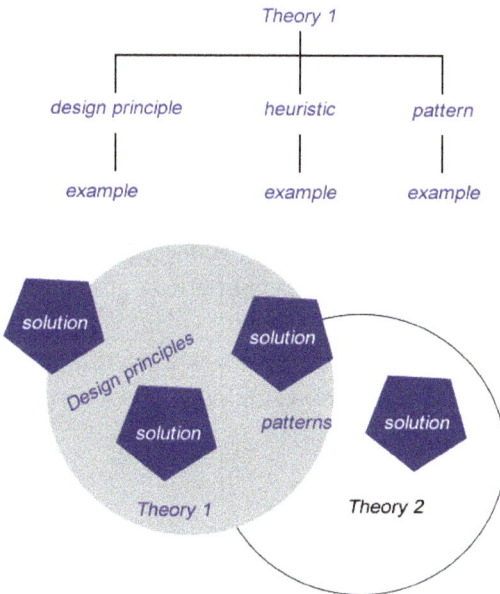

Figure 2.2. A hierarchical model of knowledge centralises theory (top image). whereas an embodied knowledge of knowledge (bottom image) centralises solutions as design alternatives that are glued together by multiple bodies of knowledge.

Van Turnhout and Smits insist that solution repertoire should be considered a performative concept. A 'repertoire' is not just a set of declarative bits of knowledge but the ability to use this knowledge to solve novel problems. For that, (aspiring) designers must practice using solutions for different knowledge problems in design (as identified by (Van Turnhout et al., 2019; van Turnhout & Losse, 2023): as a gateway to understanding problems (diagnose-oriented knowledge), a gateway to explicating design aspirations (goal-oriented knowledge), and a gateway to finding effective mechanisms in design (result-oriented knowledge).

Van Turnhout & Smits (2022) suggest that the notion of solution repertoire can be applied to the knowledge of a discipline as a whole or to a community of practitioners, not only to the individual knowledge of a designer. Van Turnhout (2023) suggests that expanding the solution repertoire of a discipline can be a primary motivation for engaging in design-based research. Novel solutions, developed in a design-based research projects have the potential to become part of the solution repertoire of the different audiences involved in these projects.

The notion of solution repertoire can thus be summarised with the following set of characteristics:

- It contains a set of concrete solutions applicable to a class of problems.
- Solutions are understood (and connected) through several bodies of knowledge (diagnose-oriented, goal-oriented and result-oriented knowledge).
- Solution Repertoire is a performative notion; it is characterised by the ability to appropriate solutions to novel situations, not just declarative knowledge of solutions.

Koen van Turnhout, Wina Smeenk, Marieke Zielhuis, Daan Andriessen

Solution Repertoire as an Evaluative Framework in ADR Projects

How can the notion of solution repertoire be adapted to be helpful as an evaluative framework for ADR projects? The central idea is that the project touches multiple practitioner communities and enriches their solution repertoire: they gain a broader set of possible solutions to refer to when addressing a novel challenge, they learn novel ways of looking at solutions they already know, or they become more fluent in coming up with new alternatives in their daily work, outside the project. We cannot specify all this impact upfront so, in contrast to the NWO model, which can be used to 'plan' for certain impact, solution repertoire needs to establish the solutions repertoires of the practice communities as the project emerges.

The first step of the approach is to identify relevant professional communities impacted by the project. These do not need to be professional designers. Educated designers may be more fluent in using solution-oriented knowledge, but one could argue that everyone engaging in some form of design applies some form of solution-oriented reasoning, thus working from a solution repertoire. However, professional communities consist largely of people who are not part of the project. The solution repertoire of an individual involved in the project may be enriched, but only if this knowledge is shared and sustained within a larger group of people, is the impact worthwhile.

The second step of the approach would be doing interviews among a representative set of people within the relevant communities to get a *qualitative feel for how stakeholders feel the research enriched their solution repertoire*. They may have experienced and learned about novel solutions, for example, because of the designs made within the project. Alternatively, they might have learned novel understandings or ways of looking at solutions they encounter daily. It is important to trace the extent to which the ideas that are sparked in the project trickle through to the daily practices of the people involved. Apart from people involved in the project, is it wise

35

to sample among those who have come in contact with the project results only indirectly, as they may have a completely different view of the importance of the project for their daily practice. As respondents may overlook some of the things they learned in the project, it can be a good idea to use the three roles of solutions we mentioned in the previous section: as a gateway to understanding problems, a gateway to explicating design aspirations, and a gateway to finding effective mechanisms in design as a basis for the interview guide. It is best to repeat this procedure several times: in the beginning of the project, because then it is much more visible what impact can be attributed to the project, but also at a later date, as some processes underlying the impact may be slow.

The third step of the approach would be to *look for evidence of wider sharing* of knowledge that originates in the project. An obvious way a project has a lasting impact is when the solutions devised in the project are implemented and used for a longer time. In such a case, people who come in contact with the solution may incorporate it into their thinking in other projects, for example. Alternatively, the solution could be shared as an inspiring example in a community of practice or incorporated in an example library (such as a pattern library or inspiration booklet). Another example of wider sharing is if the project has also delivered knowledge products such as guidelines, a card set, a toolbox, or a canvas used by the practice communities of interest. The spread of such products indicates the societal impact.

Following the three steps above should give evaluators an idea about how the project contributed in a practical way to the everyday design reasoning of the professional communities affected by the project. The idea is to do this primarily qualitatively (i.e., to trace how the solution repertoire is enriched rather than how much it is enriched) and contingent on the communities defined in the first step (i.e., it maps how different practitioner communities benefit in a different way from the project). The third step is mainly intended to give a feel for the reach (how many people are affected) and the sustainability (how long they will benefit). However, this is also intended as a rough estimate rather than a number-crunching exercise.

Example: the Dementia Simulator

The dementia simulator invites visitors to empathise with the experiences of people with dementia. Originally, it consisted of a container, in which visitors are led through a simulated experience taking the role of someone with dementia. Visitors are cleverly led into experiencing parts of the problematic daily life situations of people with dementia, including how family caregivers respond to them. Visitors claim this immersive experience is transformative: it changes how they look at people who suffer from dementia and how they evaluate their own responses to them (Figure 2.3). Rather than a physical installation, a follow-up project used Virtual Reality (VR) because VR is a more scalable technology (Spek, Sleeswijk & Smeenk, 2024). There has been a formal evaluation of the original dementia simulator (Hattink et al., 2015). This was a classic user evaluation in which visitors were interviewed about their experiences right after visiting the simulator. Although it gives a nice view of the impact of the simulator on visitors, it is limited with respect to the indirect impact.

Figure 2.3. Visitor in the dementia simulator (image from Into D'mentia).

Although the dementia simulator was not evaluated using the proposed SR-framework, we can attempt to reconstruct such an evaluation. For Step 1—identifying relevant target communities—this is feasible. The dementia simulator initially targeted two main groups: formal caregivers and informal caregivers. The core objective was to enhance their awareness of their behaviour by immersing them in the daily experiences of people with dementia. Over time, the simulator expanded to involve additional groups that interact with people with dementia in their professional roles, such as supermarket personnel, police officers, municipality officials, knowledge institutions, trainers, and health insurance providers. Designers and design researchers were integrated into the process later, particularly through the PhD research of Smeenk (2019).

Step 2—qualifying the solution repertoire of each group—is more challenging to reproduce. The formal evaluation (Hattink et al., 2015) showed that visitors gained new insights into the experience of dementia, which influenced their perception and behaviour. Since the evaluation did not account for the background of the visitors, it is challenging to determine how the simulator contributed differently to the solution repertoires of various groups. However, it is plausible that these groups encounter people with dementia in different facilities, suggesting different problems and solutions to which the simulator could potentially contribute. Informal caregivers reconsider their responses to family members with dementia, and may reconsider their caregiving routines and behaviour. In comparison, the problems police officials encounter with this target group and their action possibilities are quite different. This stipulates that the solution repertoire of each professional group involved needs to be investigated empirically. This underscores the importance of empirically investigating the solution repertoire of each professional group involved.

For Step 3—*look for evidence of wider sharing* of knowledge that originates in the project—we can say that fifteen years after the first simulator, there is much evidence of the broader reach of the simulator, which has now been translated into VR. Signs of this are the number of people visiting the simulator and the number of target groups, which we already mentioned in Step 1. Moreover, a host of knowledge products emerged from the simulator: a TED Talk, a PhD thesis, a book, a training and an e-training. The project has also been continued with a new second VR version experience about the dilemma's between formal caregiver, informal caregivers and the person with dementia. All these signify the success of the simulator, but it is still hard to establish its reach in specific target groups as these were not explicitly investigated.

Consequences of Using Solution Repertoire as an Evaluative Framework

Applied design research involves multiple practitioner groups, each with diverse and often implicit learnings. In this chapter, we propose that evaluation based on the notion of Solution Repertoire can serve as a practical and viable approach to evaluating this impact. While this is only a reconstructive case study, the example of the dementia simulator lends some face validity to the proposed evaluation approach. It appears the approach is feasible and will improve the evaluation by Hattink et al. (2015), which relied on feedback by visitors immediately after experiencing the simulator. In this final section, we compare the approach with the requirement categories identified earlier: target community, timelines, pathways, notions of knowledge and impact modes.

The proposed evaluation framework invites researchers to consider *target communities*, not merely as stakeholders in the project, but as members of broader practice communities in which these stakeholders are engaged. It stresses that these communities have a solution repertoire related to their discipline or practice, and that applied design research projects can contribute to this repertoire in unique ways for each group. This aligns with Van Turnhout and Andriessen's plea for Communities of Practice in Living Labs (2024). Like the NWO pathway approach, the proposal does not specify the target communities upfront. However, it enhances this approach by drawing attention to the knowledge buildup within these communities.

39

The proposed evaluation may capture *multiple timelines*. In direct collaborations, stakeholders' solution repertoires are expanded because novel solutions and their appreciation emerge within the project. If these are sufficiently novel and substantiated, they may also contribute to the cumulative impact of an applied design research project on practitioners' networks or other relevant communities. The evaluation, however, mainly focuses on the broader impact, which often unfolds over longer timelines.

The notion of solution repertoire does not specify the *pathways* of research impact. Other evaluation approaches, like Van Beest et al. (2022), might be more appropriate for this requirement category. The proposed evaluation framework aligns best with a *realist notion of knowledge*, acknowledging that each novel problematic context requires a unique constellation of knowledge. A solution repertoire helps in understanding and valuing situations by considering alternatives and their relative merits. This means-ends reasoning scheme could also be described as pragmatic rather than realistic. Since solution repertoire is a performative concept, the evaluation shares similarities with earlier performative evaluation proposals, like contribution mapping (Kok & Schuit, 2012). However, contribution mapping places greater emphasis on the concrete contexts in which knowledge is applied and less on the knowledge of the practitioners involved.

Regarding *impact mode*, the solution repertoire connects knowledge and learning to products and system change, suggesting that the latter are essential strongholds for the former. We 'learn' by expanding our knowledge of possible solutions to problems, their perceived benefits, and their ramifications.

Table 2.2 summarises the evaluation framework and the choices made across different requirement categories. The table suggests that our approach can address many key concerns in evaluating ADR projects, making it a promising approach. However, to validate its practicality and insightfulness, evidence must be gathered through real-life applications rather than relying solely on retrospective evaluations. This will likely be the focus of our next steps.

40

Category	Contribution of Solution Repertoire based evaluation
Target Community	Need to be defined by the evaluators in an explicit step
Timelines	Is mostly targeted to indirect impact on longer timelines
Pathways	Are not part of the evaluation
Type of knowledge	Is realistic or pragmatic
Impact modes	Products and system change yield knowledge and learning

Table 2.2. Solution Repertoire concerning aspects of impact.

Key Takeaways

- ADR projects have both direct and indirect impact on society. A practical approach is needed to identify and evaluate these impacts.
- We propose that ADR projects can be evaluated by identifying practitioner groups impacted by them and evaluating how the project contributed to their solution repertoire: their knowledge and understanding of solutions relating to the class of problems they face.
- A reconstruction of the evaluation of the dementia simulator suggests such an evaluation addresses many key concerns with existing evaluation approaches and is an improvement on more proximate evaluation impact approaches.

References

- Arlitt, R. M., Stone, R. B., & Tumer, I. Y. (2015). Impacts of function-related research on education and industry. In Chakrabarti A., & Lindemann, U. (eds.) Impact of Design Research on Industrial Practice: Tools, Technology, and Training (pp. 77–99). Cham: Springer International Publishing.
- Chakrabarti, A., & Lindemann, U. (2016). Impact of design research on industrial practice. Impact of Design Research on Industrial Practice.

• Coombs, S. K., & Meijer, I. (2021). Towards evaluating the research impact made by universities of applied sciences. Science and public policy, 48(2), 226–234.

• Greven, K., & Andriessen, D. (2019). Practice-based research impact model for evaluation: PRIME. In EAIR Conference (Vol. 2019, pp. 1–11).

• Hattink, B. J., Meiland, F. J., Campman, C. A., Rietsema, J., Sitskoorn, M., & Dröes, R. M. (2015). Experiencing dementia: evaluation of Into D'mentia. Tijdschrift voor Gerontologie en Geriatrie, 46, 262–281.

• Into D'mentia (n.d.). https://intodmentia.nl/

• Kok, M. O., & Schuit, A. J. (2012). Contribution mapping: a method for mapping the contribution of research to enhance its impact. Health research policy and systems, 10, 1–16.

• Lawson B. R. Cognitive strategies in architectural design. Ergonomics, 22, 1979, 59–68.

• NWO (n.d.). https://www.nwo.nl/en/impact-plan-approach, consulted 6–6–2024.

• Pasman G. Designing with precedents (Unpublished doctoral dissertation), Delft University of Technology, Delft, NL, 2003.

• Smeenk, W. (2019). Navigating empathy: empathic formation in co-design.

• Smeenk, W. (2022) Societal Impact Design: empathic and systemic co-design as a driver for change. In: Joore, P., Stompff, G., Eijnde, J. van den (eds). Applied Design Research: A Mosaic of 22 Examples, Experiences and Interpretations Focussing on Bridging the Gap between Practice and Academics. CRC Press. P147–155.

• Smeenk, W., Sturm, J., & Eggen, B. (2018). Empathic handover: how would you feel? Handing over dementia experiences and feelings in empathic co-design. CoDesign, 14(4), 259–274.

• Smeenk, W., Zielhuis, M. & van Turnhout, K. (2024). Bridging the research-practice gap: Understanding the knowledge exchange between design research and social design practices. *Journal of Engineering Design, 1–21.*

- Spaapen, J., & Van Drooge, L. (2011). Introducing 'productive interactions' in social impact assessment. *Research Evaluation, 20*(3), 211–218.
- Van Turnhout, K. (2023). Ontwerpen als onderzoek. In K. van Turnhout, D. Andriessen, & P. Cremers (Eds.), *Handboek ontwerpgericht wetenschappelijk onderzoek: Ontwerpend onderzoeken in sociale contexten* (pp. 24–37). Boom Uitgevers.
- Van Beest, W., Boon, W. P., Andriessen, D., Pol, H., van der Veen, G., & Moors, E. H. (2022). A research pathway model for evaluating the implementation of practice-based research: The case of self-management health innovations. *Research Evaluation, 31*(1), 24–48.
- Van Turnhout, K., & Andriessen, D. (2023). Experimenting with novel knowledge: A plea for communities of practice. In P. Joore, A. Overdiek, W. Smeenk, & K. van Turnhout (Eds.), *Applied design research in living labs and other experimental learning and innovation environments* (pp. 148–165). CRC Press.
- Van Turnhout, K., Jacobs, M., Losse, M., van der Geest, T., & Bakker, R. (2019). A practical take on theory in HCI. *White paper.*
- Van Turnhout, K., & Losse, M. (2023). Kennisfuncties in ontwerpgericht onderzoek. In K. van Turnhout, D. Andriessen, & P. Cremers (Eds.), *Handboek ontwerpgericht wetenschappelijk onderzoek: Ontwerpend onderzoeken in sociale contexten* (pp. 233–248). Boom Uitgevers.
- Van Turnhout, K., & Smits, A. (2021). Solution repertoire. In *DS 110: Proceedings of the 23rd International Conference on Engineering and Product Design Education (E&PDE 2021)*, VIA Design, VIA University in Herning, Denmark, 9–10 September 2021.
- Van Turnhout, K., & Smits, A. (2022). Radio Dabanga: Applied design research in human experience & media design. In *Applied design research* (pp. 33–42). CRC Press.
- Weiss, C. H. (1997). How can theory-based evaluation make greater headway? *Evaluation Review, 21*(4), 501–524.
- Zielhuis, M. R. P. (2023). *Considering design practice: The underutilized opportunities in collaborative design research projects for learning by design professionals.*

43

Roelof A.J. de Vries, Judith Weda,
Shakila Shayan, Koen van Turnhout

3. The Impact of Technology Maturity on Determining Urgent Research Questions

DOI: 10.1201/9781003609766

Introduction

In science, a common understanding is that a good research question should be capable of producing results that are valuable, useful, and achievable. However, determining what makes a good research question is not a trivial undertaking. In formulating projects that aim for societal impact, one way to look at 'good' research questions is by asking 'urgent' questions or aiming for the most 'urgent' goals (Stojanovic, 2023). Scoping research projects to these urgent questions or goals, however, is a difficult component of the research process as it involves many practical and theoretical considerations as well as many uncertainties within and beyond the project. This is true for the research process in general, but in particular for design research, as it has the ambition to scope the research goals in such a way as to play a proactive role in societal transitions. As such, it is always dealing with moving targets: ever-changing wants, needs, and opportunities in society. This contribution focuses on identifying the maturity of a particular technology to be used, to help formulate 'fitting' design research projects about this technology. In addition, this chapter articulates different trajectories of technology in design research in terms of its naturalization, building on the Pyramid of Technology by Koert van Mensvoort (2013).

In this chapter, we first analyze the evolution of one particular technology, Virtual Reality (VR) Technology, through the lens of the Pyramid of Technology (see Figure 3.1) by Koert van Mensvoort (2013). Second, we map existing research questions of our research group (Human Experience & Media Design at the University of Applied Sciences Utrecht) to the pyramid.

With the Pyramid of Technology, Koert van Mensvoort introduces a model that describes how emerging technologies 'naturalize' over time. The model or pyramid consists of seven levels that explain how technology functions in our lives. Of the seven levels of the pyramid, the lowest level of technology described is the envisioned level. The *envisioned* level of technology is a hypothetical stage where the technology does not actually exist yet, but is being dreamed up in the mind of the inventor, for example, How could cold fusion solve the world's energy crisis? The *operational* level describes a technology of the proof-of-concept level where the technology has been made but is not widely applied and usually only exists as a prototype; for example, How would brain-computer interface implants influence daily interactions? The *applied* level describes technology that has moved out of the concept phase and into the early consumer stage. People can actually run into this technology now, although it is not yet widespread throughout society; for example, How can we use facial recognition software for societal good? The *accepted* level does describe technology that is now widespread throughout at least some societies, although there are technologies that are not yet accepted across all cultural or social dimensions; for example, How can we design for users to have more influence over their recommended content? The *vital* level describes technologies that have reached a ubiquitousness or pervasiveness that has many (but not all) people dependent on the technology; for example, How has the World Wide Web shaped modern society? The *invisible* level describes technology that we no longer notice. This level is not often attained by any technology and is reserved solely for technologies that we do not usually consider technology anymore, such as writing. An associated research question could be How has writing shaped our relation to story-telling? The last level is the *naturalized* level that describes technologies that we consider to be part of human nature. Long ago these technologies might have been novel, but in our current day and age these technologies are experienced as second nature, such as cooking or language. An associated research question could be How has language shaped our species short-term evolution? The vital, invisible, and naturalized levels are usually not part of the scope of any design research projects.

The technologies that we research in a fundamental as well as applied academic context fit into the lower levels of the pyramid. The pyramid can be used to understand differences between fundamental and applied design research. Generally speaking, the technologies that more *fundamental* design research considers are usually either in the envisioned or operational levels of the pyramid. This type of research usually focuses on counterfactuals or what-if questions: 'What if we could design a technology that solves problem X?', 'What if we could make a technology that does

Y?'. In this academic context, dreams are dreamed and prototypes are prototyped, but there is no real need or incentive to go beyond this stage, the fundamental concept has been proven, or at least the groundwork has been laid, and the day is called. In contrast, the technologies that *applied* design research considers in projects are also in the envisioned and operational levels, but even more so in the applied, accepted, and perhaps sometimes vital levels of the pyramid. Specifically in the research context of applied design projects, all five lower levels (envisioned as vital) are possible levels of naturalization that are studied, sometimes more than one within a project, as we will show in this chapter. By using the pyramid as a lens to look at the evolution of one specific technology and at a wide variety of research questions, we can reflect on and justify the type of question or aim that is most suitable to the given specific technology maturity within and between the problem contexts. For example, virtual reality technology for children with a specific disability could still be at an operational level, while virtual reality technology for specific types of gamers could be at an accepted level. These contexts ask for completely different 'urgent' questions or aims, while the used or proposed technology is identical.

In sum, we argue that the model of the seven levels of technology maturity can serve as a useful lens to analyze and delineate research projects, which, in turn, can help reflect on but also justify the type of research question or aim that is most suitable to ask given the specific technology maturity within the problem context.

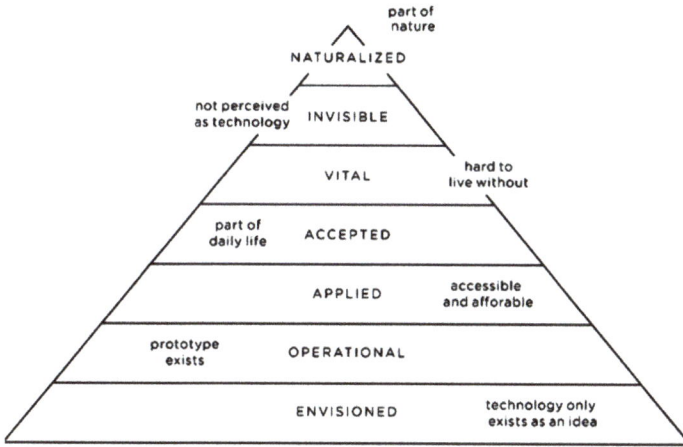

Figure 3.1. Koert van Mensvoort's Pyramid of Technology (van Mensvoort, 2013).

Example: The Evolution of Virtual Reality Technology

To illustrate the use of the Pyramid of Technology as a lens, we look at the evolution of VR technology as described by Kenwright (2019). In this chapter, Kenwright (2019) specifically delineates the evolution (or naturalization) of VR technology in four noteworthy periods, or epochs (see Figure 3.2).

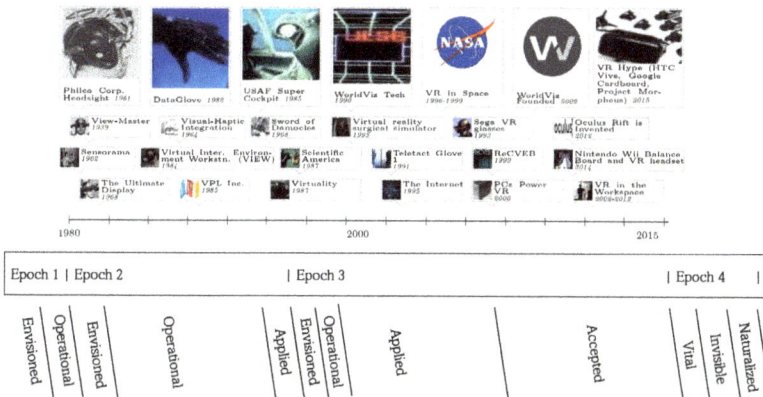

Figure 3.2. Adapted illustration of VR technology of the years by Kenwright (2019) matched with levels of naturalization of the technology.

Epoch 1 is described as the "early years" (Kenwright, 2019, p. 5), and consisted of works of VR technology that were not digital, such as the View-Master and the Sensorama (see the adapted Figure 3.2).Words used to describe this period such as "creativity", "imagination", "artistic initiative" suggest that this technology resided mostly on an *envisioned* level of the naturalization of technology. The questions explored through these technologies revolve mostly around what *virtual* technologies *could* look like, they are mostly counterfactuals or what if questions. Moreover, many prototypes have also been made in these contexts, which would also suggest an *operational* level of technology, where the designs function as a proof of concept for perhaps a larger vision for the future role of VR technology.

Epoch 2 introduced the first visions for *digital* VR technology, which is the implied meaning of VR technology in the contemporary use of the word. This epoch represents the first steps of Human-Computer Interaction embracing the idea of digital virtual reality. This new perspective on VR also introduced the possibility of wearable VR, with the first consumer VR headsets made. The technology in this epoch is described as having "all sorts of amazing possibilities" (Kenwright, 2019, p. 6). The change of technology to a more digitally oriented version of VR implies that a quick *re-envisioned* level has happened, and perhaps more importantly, a longer *operational* level of the naturalization of technology happened where a number of promising and innovative prototypes were made. This epoch ends in a very short *applied* level of technology, where the first consumer products were made, but never gained traction.

Epoch 3 represents our current period of exposure to and interaction with VR technology. Due to, amongst other things, technological advancements, VR technology resurged and is a somewhat commonplace technology in the current day and age. Words used to describe this epoch of VR technology include "virtual environment" and "immersion", which suggest a much more *applied* level of technology. The current state of the art is much more invested in particular application areas than in other epochs, and research is therefore concerning itself with much more applied questions, such as VR for exposure therapy (Krijn et al., 2004), or VR for training astronauts (Garcia et al., 2020).

Epoch 4 envisions the future of VR technology. It has not yet come to pass and as such is uncertain, but predictions are made in academia as well as science fiction.Most predictions follow the order of the pyramid of technology, where an ever increasing naturalization of technology is imagined. The *vital* level of VR technology imagines an increasing dependency on VR for all kinds of training and entertainment. The *invisible* level of VR technology imagines a ubiquitousness and pervasiveness of technology that much of reality *could* be virtual (e.g. all walls of your house have a double function as screens, and can therefore present you a virtual reality seamlessly). Whether the *naturalized* level is achievable and imaginable is perhaps a question of definition, because naturalized in the sense of 'part of human nature' seems antithetical to virtual in the sense of 'not real'. In terms of questions, it is interesting to notice that the questions become *envisioning* again, but in this case in terms of if and how VR technology could play a role at a *vital* level.

When applying the dimensions of naturalization of technology to the evolution of VR technology as described and delineated by the aforementioned epochs, it is interesting to notice that these levels of naturalization of technology fit well, but are also not as linear as suggested. To gain some additional perspective into this non-linearity, we apply these levels of naturalization of technology to diverse present-day research questions from our own research group.

Workshop on Relevant Research Questions in the Pyramid of Technology

In a workshop with our research group of twelve design researchers we explored if and how our research questions map to the specific levels of naturalization of technology. The design researchers were asked to collect the research questions they have recently worked on, they are currently working on, or are planning to work on shortly (see Table 3.1).

On a general level, it is immediately clear that none of the research questions were mapped to the *vital*, *invisible*, or *naturalized* levels of technology naturalization. This is both a reflection on the type of technology we are generally interested in in design research, as well as the types of questions we ask about this technology.

The *envisioned* and *operational* levels are also not broadly represented in our mapping of research questions. The questions that do map to the envisioned level are not necessarily mapped there because the technology they are about is at an envisioned level, but because the questions are aspirational and broad, leaving the technology open and in that sense, envisioned, but not *necessarily* in the sense that it does not exist yet. The questions that do map to the operational level are not necessarily mapped there because the technology that is being developed or used is solely available at the operational level, but because the questions aim for the development of a certain technology only to the extent of functioning at the operational level, e.g. prototypes.

Most questions are mapped to *applied* and *accepted* levels of naturalization. The questions mapped to the applied level are of a quite diverse nature, but a common theme in these questions is that they do cover certain aspects of the usability of user experience that can be improved. For the questions mapped to the accepted level it is interesting to note that a common theme is the application of this accepted level of technology in a new setting where it can hopefully be useful as well.

Overall, the mapping of the research questions of our applied design researchers to naturalization levels are not surprising. Most questions focus on the *applied* and *accepted* levels, which seems appropriate for *applied* design research.

The technologies that we look at in our design research are in the lower and intermediate levels of naturalization, as shown by our workshop on mapping applied design research questions to naturalization levels. This understanding of the naturalization of technologies can help to understand and perhaps even signify the difference between fundamental and applied design research. Fundamental design research usually concerns itself with the *envisioned* or *operational* technology, 'counterfactuals' and 'what-ifs' are core questions for fundamental design research. In contrast, applied design research concerns itself with technology in a very practical manner, looking to investigate and influence the immediate impact the technology is having, which is of chief concern in the *applied*, *accepted*, and *vital* levels of technology. The invisible and naturalized levels of technology are usually out of scope for applied design research, perhaps because we do not notice it anymore and therefore do not question the impact of the technology.

We looked at the evolution of a particular technology, VR technology, through the lens of the Pyramid of Technology (van Mensvoort, 2013), see Figure 3.1. The pyramid specifies seven levels to the naturalization of technology in societies. We argue that looking at your design project through the lens of these seven levels can help determine the right 'urgency' for your question. However, the seven levels are not always straightforward or linear, as we reflected on in a workshop with twelve design researchers. During the discussion part of this workshop and in the plenary mapping of research questions to naturalization levels it became apparent that it is not so trivial to decide the naturalization level of the technology the research question is focused on. This is because—as opposed to the social sciences—generally speaking, design research is not focused on a description or explanation of what is, but on a plan or solution for what could be, will be, or even ought to be. From that perspective, design research is about possible worlds. Reflecting on the naturalization levels of technology, the research questions that design researchers ask are about the transitions *between* naturalization levels.

This perspective is represented in Figure 3.3. In this figure we reimagine a way to map applied design research questions to naturalization levels of technology in terms of a 'trajectory' that we aim for with this technology. For this new trajectory that we imagine throughout the pyramid, the starting point is to assess the general naturalization level of the technology you are starting from, and display the transition you aim for in terms of re-application of this technology. For example, the general naturalization level of the technology can be at an accepted level (e.g. robots in warehouses), and the transition you aim for is to also get the technology at this same or perhaps a lower level or higher level of naturalization in a new context (e.g. robots in a hospital setting). This conceptualization of the transition that applied design research tries to generally achieve works for most of our research questions. Most of our research questions aim to reuse a technology and its associated knowledge in a new context to achieve transitions in that context. This conceptualization of the transition that applied design research aims for, combined with the naturalization level of the technology, can help in shaping or scoping the research project by envisioning the effort needed to achieve the imagined trajectory, but can also help determine the urgency of research questions. This conceptualization can serve as a lens for looking at the research project that helps reflect on whether the technology is scoped to the right level of naturalization in that particular context to have the 'urgency' or the societal impact the research question aims for. Too 'bold' or 'adventurous' levels of naturalization for a particular technology might sound valuable but in practice might not be useful (yet) or achievable, while too 'undemanding' or 'elementary' levels of naturalization for a particular technology might be achievable, but will not result in much value or usefulness for society.

In the Algorithmic Affordances (Smits & van Turnhout, 2023) project with Recommender System technology (see Table 3.1, RQ17), we want to improve the quality of the implementation, and make the existing design choices and their underlying knowledge more visible and usable, so it takes an accepted technology and brings it back to an applied level. In the Social Robot project, we want to take an applied technology in a related context to a new context (see Table 3.1, RQ8 and RQ9), so the transition is from applied 'down' to envisioned, and 'up' to operational in a new setting. Still, not all our research questions fit this new concept of combining trajectory plus naturalization levels of technology. In the workshop, two research questions did not specify the technology beforehand, but rather the methodology to be used. In this case, it is impossible to map the research question to the naturalization levels, because the technology to be designed for is not determined yet. This can be considered a different class of research questions that could benefit from a different conceptualization of what are urgent questions in methodological terms.

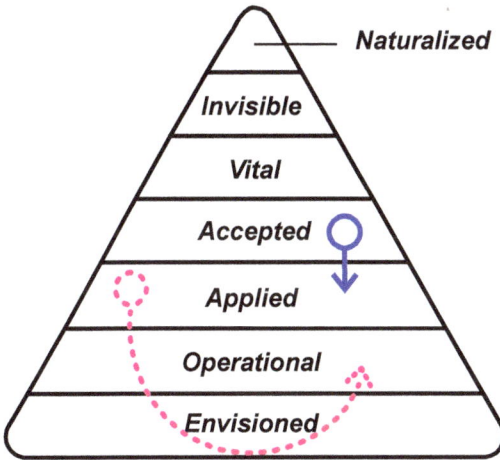

Figure 3.3. The pyramid of technology with example trajectories in pink. Indicating the social robot technology (RQ8 and RQ9 in Table 3.1) already applied in a specific context with a circle, which then moves along a dashed line to the operational level in a new context. The recommender system (RQ19 in Table 3.1) starts at the accepted level and moves one step down to the applied level.

RQ	Envisioned	Technology
1	How can intelligibility and pronunciation of Dutch by second-language learners be improved and how can existing knowledge and ability in the disconnected second-language learners work domain be combined and/or gathered?	Cardset
2	What do children, parents and healthcare professionals need for better quality of life for a child in palliative care?	Data tracking app
3	How can we accelerate the sustainability transition by means of Immersive Design methods?	Undetermined
4	How do underlying assumptions, norms, and dynamics of political emotions interact with the design of Digital Democratic Instruments to contribute to the participation paradox?	Online participation platform
	Operational	
5	To what extent do users of the proposed Digital Democratic Instrument-designs experience these designs as supporting agency (in real world settings)?	Online participation platform
	Applied	
6	In which way can we strengthen the connection and trust between journalism and citizens by AI-driven conversational technologies?	Conversational AI
7	How can we evoke emotions with biodiversity data by the means of Immersive Design in order to promote behavior change with a positive effect on the natural environment and sustainability?	Undetermined
8	How does the design and use of the droomrobot connect with the broad practice in hospitals?	Social robot
9	How should the droomrobot's physical shape and interaction be designed in order to effectively apply guided fantasy for 4- to 12 year old children in the broad context of hospitals?	Social robot
10	Does presenting course content via the gamified 'InPraktijk' app generate more interest in physical therapy students than generic content delivery?	Learning management system
11	How can we integrate curricular and experiential knowledge in explainable AI systems for the financial technology sector?	Explainable AI
12	How can Digital Democratic Instruments be designed to positively influence the capacity and exercise of both individual and collective agency?	Online participation platform

13	What barriers in the current design of Digital Democratic Instruments hinder the effective expression of political emotions?	Online participation platform
	Accepted	
14	What algorithmic affordances are there?	Recommender system
15	How can providers of cardiac rehabilitation use (exercise) data to guide the content and timing of treatment in order to provide meaningful care?	Dashboard
16	Does a data tracking app to track self-reported symptoms of children in palliative care help with better symptom recognition and comfort?	Data tracking app
17	How can we support designers with responsible interaction design for recommender systems so that they can contribute to the development of responsible AI?	Recommender system
18	What should a learning management system offering educational content to pediatric physiotherapists consist of?	Learning management system
19	What characteristics of existing citizen participation platforms are inclusive and exclusive?	Online participation platform
	Vital	
	-	-
	Invisible	
	-	-
	Naturalized	
	-	-

Table 3.1. A selection of results from the workshop with design researchers, where we mapped current research questions on the pyramid of technology. The technology matches the research question and not necessarily the maturity level.

Conclusion

In this chapter, we tried to illuminate some of the considerations and characteristics that play a role in formulating 'urgent' questions for design research projects. Asking 'urgent' questions is important for design research, because—as designers—we want to play a proactive role in societal transitions and imagine futures worth wanting. This is particularly difficult because societal transitions are about the ever-changing wants, needs, and opportunities in society, and as such designers are always dealing with moving targets. To this end, we made the case that, in part and with the

57

acknowledgement that a great many factors contribute, looking at the trajectory of a question can help define a valuable, useful, and achievable design research project with societal impact. The urgency of questions can be partially substantiated by the trajectory and the naturalization level of the technology that is being researched.

Key Takeaways

As designers, we want to play a proactive role in societal transitions and have impact. To this end, we feel that asking 'urgent' questions for your design research projects is crucial. We show that:

- Design research questions are typically about achieving transitions between naturalization levels of technology, which we call the 'trajectory of the technology'.
- Considering the trajectory of the technology you are researching can help in formulating and delineating relevant research.
- Considering the trajectory of the technology you are researching can help in understanding what transition of technology you are aiming for in your design research.

References

• Garcia, A. D., Schlueter, J., & Paddock, E. (2020). Training astronauts using hardware-in-the-loop simulations and virtual reality. In *Proceedings of the AIAA Scitech 2020 Forum* (p. 0167). https://doi.org/10.2514/6.2020-0167

• Kenwright, B. (2019). *Virtual reality: Where have we been? Where are we now? And where are we going?* Preprints. https://doi.org/10.20944/preprints201907.0130.v1

• Krijn, M., Emmelkamp, P. M., Olafsson, R. P., & Biemond, R. (2004). Virtual reality exposure therapy of anxiety disorders: A review. *Clinical Psychology Review, 24*(3), 259–281.

• Smits, A., & van Turnhout, K. (2023). Towards a practice-led research agenda for user interface design of recommender systems. In *IFIP Conference on Human-Computer Interaction* (pp. 170–190). Springer Nature Switzerland.

• Stojanovic, M. (2023). Pursuitworthiness in urgent research: Lessons on well-ordered science from sustainability science. *Studies in History and Philosophy of Science, 98*, 49–61.

• Van Mensvoort, K. M. (2013). *Pyramid of technology: How technology becomes nature in seven steps*. Eindhoven University of Technology.

Gijs Terlouw, Lars Veldmeijer

4. Co-designing Potential Boundary Objects to Open Patient Perspectives and Redressing Power Imbalances in Mental Healthcare Design Studies

DOI: 10.1201/9781003609766

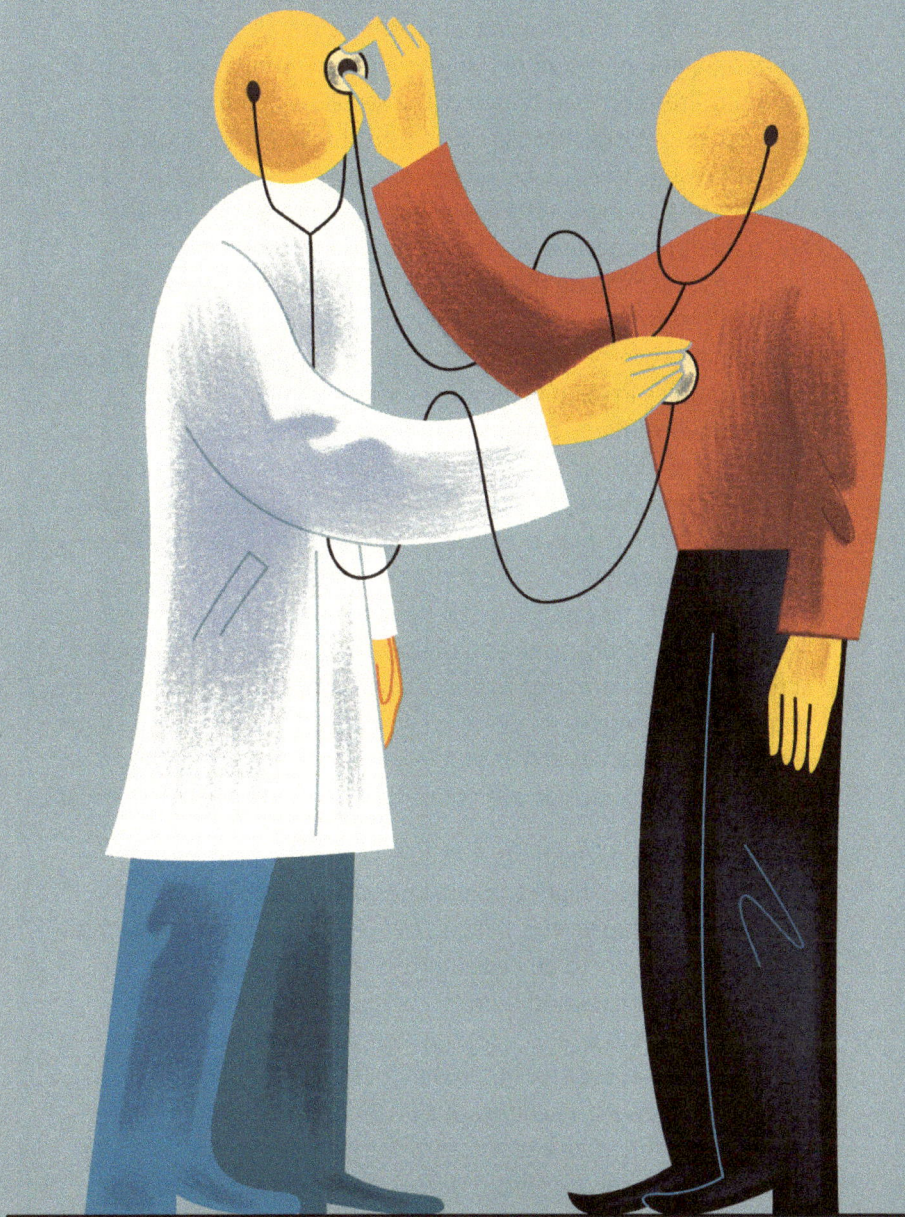

Introduction

In 2023, influential design theorist Emeritus Professor Nigel Cross highlighted the immense growth of publications on design thinking (Cross, 2023). Similarly, the use of applied design research in mental healthcare is growing, and terms such as design thinking, co-design, and participatory design are increasing in the medical scientific literature. On PubMed, the results on co-design and participatory design have grown from 72 results two decades ago to 1847 results in 2023. Co-design and participatory design have the potential to more effectively involve patients and harness the patient perspective as a key component in developing solutions and shaping the future of healthcare. However, two reviews (Veldmeijer et al., 2023; Brothersdale et al., 2024) who studied the current role of patients in co-design and participatory design studies in (digital) mental healthcare conclude that patient involvement remains minimal. In most studies, patients' participation does not go beyond responding to pre-developed scenarios and concepts. In only a few studies has the patient been a partner in the design process and allowed to co-decide on the direction of a solution. The patient perspective, which requires more in-depth research into the needs and experiences of clients and patients, rarely receives a place in those studies. Especially at a time when societal trends are calling for more participation and a more prominent role for patients in addressing their mental health issues, it is remarkable that studies applying a co-design approach fail to effectively include patient perspectives in the design process.

Medical research traditions and culture are dominant in mental healthcare, which might explain why patient participation on a more partnership level is difficult to achieve in practice. Blandford et al. (2017) reflected on seven differences on a cultural level between healthcare and human-computer interaction design, while Groeneveld et al. (2018) described several challenges a designer may encounter in the healthcare sector. One of the major differences between healthcare culture and design culture lies in the definition of an expert: from the medical perspective, it is always the specialist or professional, whereas from the design perspective, it is always the end user. The perspective of the specialist or professional as the expert dominates the medical sector, which is even further reinforced by the hierarchical structures within healthcare organizations. As a result, when terms like co-design and participatory are applied in mental health design studies, it often leads to a scenario where the patient adopts the dominant perspective of the specialist or professional. Consequently, the patient mainly becomes someone who responds to ideas and concepts that have already been shaped from the specialist's perspective. With movements like Positive Health (Huber et al., 2011) and the recovery movement, where the inclusion of the patient's perspective is essential, a shift in power imbalances in design studies seems increasingly desirable. This does not mean a shift from one perspective to another, but rather a movement where both perspectives coexist more equally.

Specialists and patients who collaborate in shaping future mental healthcare may encounter boundaries due to cultural differences and the strong hierarchy in mental healthcare. People who share the same community easily understand each other and share knowledge (Lave & Wenger, 1991), but sharing knowledge across boundaries can be challenging (Carlile, 2002). Designers and design approaches hold the potential to overcome these boundaries. Stompff (2013) describes the boundary-spanning capabilities of designers in overcoming boundaries, while Island et al. (2017, 2019) highlight the potential of boundary objects to bridge boundaries in healthcare. Boundary objects were first described by Star and Greisemer (1989) as objects that allow different groups to collaborate without consensus on goals, needs, and outcomes. In addition to the potential of boundary objects, Akkerman and Bakker (2011) described the dialogical learning mechanisms boundary objects can enact across different stakeholders in education. These core learning mechanisms are identification, coordination, reflection, and transformation. Reflection involves expanding perspectives through perspective-taking and perspective-making. In healthcare, when the reflection mechanism is triggered, it increases the chances of successful transitions and transformations (Terlouw et al., 2022; Kajamaa, 2011; Sajtos et al., 2018; Jensen & Kushniruk, 2016). Developing and designing potential boundary objects from the view that multiple

63

stakeholders can work together without consensus, without giving up their own viewpoint or letting another dominate, can provide patients more space to play a bigger role in co-design processes in mental health. In this chapter, we describe two cases in which we attempted to shape potential boundary objects through co-design to open up and address the patient perspective.

AscapeD

In the first applied design research study, we developed the serious game AScapeD, mainly focused on children with the classification autism spectrum disorder (ASD). Through the application of different potential boundary objects during the design research process, we gained a much richer picture of the problem- and solution context from both the medical and the patient perspective. Initially, the project's objective was to create a serious game to improve the social skills of children with ASD. However, the assignment was formulated from the perspective of specialists only; by pursuing the development of a potential boundary object, other perspectives opened up.

Figure 4.1. Persona description of Joris.

Through a co-design design approach (Terlouw et al., 2020) we applied different design methods, for instance, a persona (Figure 4.1). We found out that children with ASD don't face many problems with their social skills, they mainly want to be part of the group and interact with their peers. The biggest problem the children face in daily life is that they do not fit in. Where parents and professionals think that developing social skills through social skill training is essential to pursue the goal of fitting in, children themselves do not connect those dots. The participating children in the study regularly turned out to be victims of physical or verbal violence by their peers. The children did not blame the problematic social situations they encountered at school on their lack of social skills, as the goals of the social skills training suggests.

By opening up this perspective, we were able to set more tailored goals for the serious game; from a therapeutic perspective, aspects of social skill development, like turn-taking, cooperation, joint attention, and vocalization, were included. From the children's perspective, forming more friendships, connecting with peers, experiencing more peer acceptance, and enacting mutually enjoyed activities were included (Terlouw et al., 2021). Based on these goals, we started the iterative development of AscapeD, where we developed and tested different prototypes (Figures 4.2 and 4.3) with the objectives from both perspectives. During the project, through tests, we also expanded the perspectives to other relevant stakeholders, like the teachers at school and the peers of the children with ASD.

Figure 4.2. Paper prototype of AScapeD, supplemented with materials from an existing escape game.

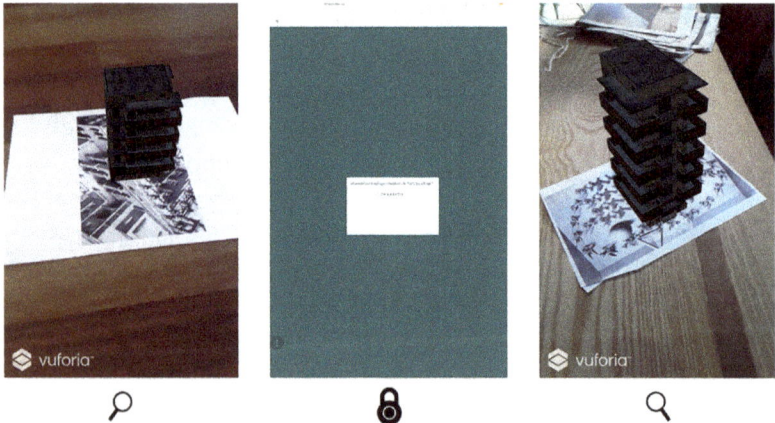

Figure 4.3. Prototype of AScapeD, a simple AR-based game.

Eventually, AScapeD was developed (Figures 4.4 and 4.5). Within this serious game, the players work as detectives in the case of a missing girl named Charlotta. The players are introduced into the case with a short story and placed in the girl's room. All the rooms have the same appearance; however, each room's time is different for each player; therefore, various objects are placed in other places. Players advance through the game by solving riddles and bringing together information from different rooms through face-to-face communication.

Figure 4.4. Screenshot of AScapeD.

Figure 4.5. Three-player setup of AScapeD.

By developing the serious game as a potential boundary object, the process did not result in a new tool for a specific stakeholder system. The process resulted in a mediating tool that contributes to different involved activity system objectives without trying to reach a consensus between them. AScapeD adapts to different stakeholders' local needs and constraints and obtains different meanings for different activity systems.

67

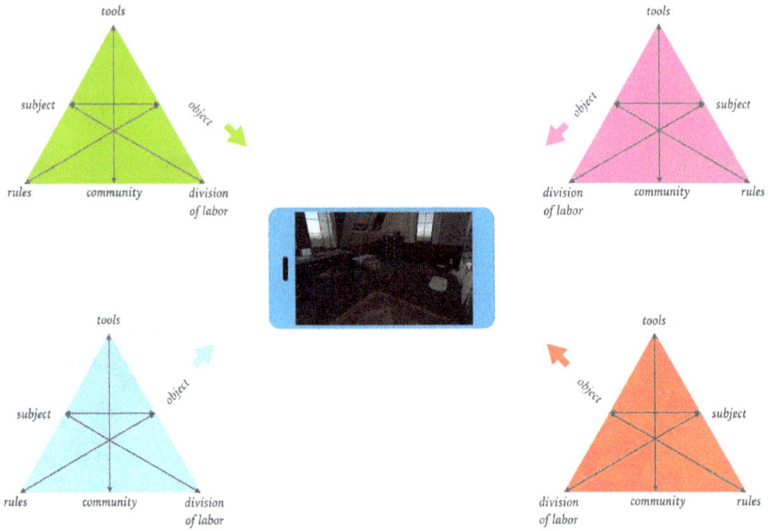

Figure 4.6. AscapeD as a boundary object between different stakeholder systems.

In Picture Approach

In the second applied design research study, we focused on the diagnostic process of people with altered perceptions. People with altered perceptions give excessive meaning to the world—they are in a state of hyper-meaning. This state of hyper-meaning is also known as psychosis. Lived experience research shows psychosis is experienced as a painful and upsetting existential experience—alien to our usual patterns of life and simultaneously enigmatic and human (Fusar-Poli et al., 2022). Every individual has a different experience of psychosis. To help people with mental health problems in mental health care, such as psychosis, professionals conduct diagnostics. Diagnostics enable the professional to understand both the mental suffering and, ideally, the person who suffers and the context behind the suffering. However, from a phenomenological viewpoint, this diagnostic process is methodologically and epistemically limited in making people's idiographic experiences of mental health issues tangible (Veldmeijer et al., 2023; Veldmeijer et al., 2024).

In the case of patients who suffer psychosis this is even more challenging. It is difficult for professionals to fully understand the experience of psychosis without having experienced it themselves. Even when professionals possess experiential knowledge of psychosis, each experience is still filled with personal meaning in the context of an individual's life story (Corstens & Longden, 2013; Ritunnano et al., 2022), making each experience different. However, trying to understand the individual experience of psychosis is important to providing person-centered care. An essential question in this process is: how can a patient be assisted in 'capturing' such an experience, as it is so sensory and full of personal meaning? The current diagnostic procedure falls short in providing adequate methods that allow individuals with psychosis to make their experiences tangible while also allowing the professional to understand the idiographic experience. Additionally, there is the issue of a tendency to 'reach agreement' on care needs, but how can there ever be agreement if the patient and professional do not fully understand what is being experienced?

To prevent this urge to search for agreement, we used the concept of boundary objects within in this design project to bring both perspectives to the table. Specifically, we focused on the social effect that the bridging function that a potential boundary object can have in the patient-professional alliance in co-understanding psychosis. The aim was to design a meaningful artefact from multiple perspectives without requiring the different stakeholders—professionals and patients—to agree on what the artefact exactly is.

To give people with lived experience of psychosis a central role as architects of the artefact, we started the design phase with generative design workshops. This follows the latest developments in mental health care innovation, where people's lived experiences are increasingly being involved as partners in design projects to discover new perspectives to reframe the problem or to develop novel solutions (Nakarada-Kordic et al., 2017; Marshall et al., 2024). In these design workshops, individuals with lived experience and professionals, under the guidance of students and the supervision of qualified researchers and clinicians, developed concepts to make

69

sense of individual psychosis experiences. The central question in the first design activity was: What would have helped you to give meaning to your psychosis? In the generative design workshops, we emphasized that the participants did not have to agree on the innovations exactly, but that developed concepts had to be helpful for all parties involved. The results varied widely (Figure 4.7).

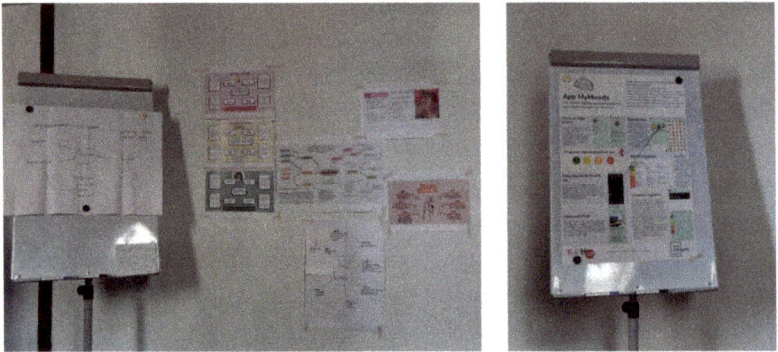

Figure 4.7. Impression of two concepts designed by people with lived experience of psychosis, professionals and students.

Although the developed concepts were all different in content—which is not surprising given that they were based on the participant's unique individual experiences of psychosis—there were also commonalities. For instance, all the concepts focused on visualization processes rather than giving words to the experience. Even more interesting, when discussing the concepts, we learned that the rationale behind the concepts shared notable similarities: all the concepts served two essential purposes. For the professional, the concepts functioned as 'translation tools'. For the people with lived experience, the concepts functioned as 'exploration tools' that encouraged self-research and pursuing personal strength. As such, the concepts enabled the professional to understand the individual better because the experience was translated into an understandable visual language. Simultaneously, it offered the individuals with a psychosis a tool to make sense of altered perceptions and hyper-meaningful experiences in the context of their life story.

Interestingly, the participants discussed and analyzed in a design workshop why they believed the developed concepts shared these similarities. Based on their lived experience, they noted that current methods are primarily language-based and mainly focus on the needs of the professional to understand the patient, while their own developed concepts promoted patients' agency to explain their experience to the professional. As a result of this design workshop, the design team decided to continue iterating based on these principles. After three iterations, this led to the concept of the 'In Pictures Approach (IPA)', a visualization-based tool centered on creativity and self-exploration in the diagnostic process, addressing both the challenges and strengths of the patient, as well as the challenges for professionals to understand the individual experience of psychosis (Figure 4.8).

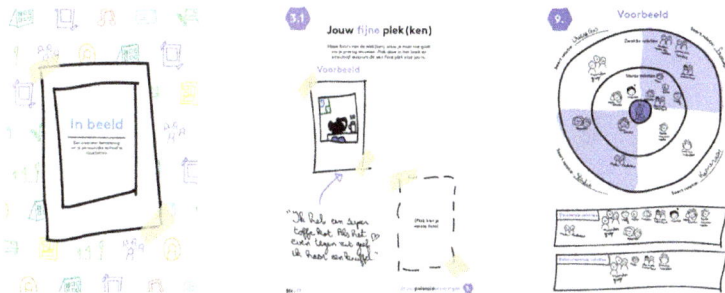

Figure 4.8. Results design workshops focused on the product—In Pictures Approach.

After the expert group believed the concept of IPA was promising for patients and professionals, a design session was planned to focus on how and where to implement IPA within the current diagnostic process. This session was primarily aimed at integrating the approach into the system and was conducted with two groups of professionals and individuals with a lived experience. The current and desired care journeys were visualized during this workshop, and opportunities, limitations, and prerequisites for using IPA in the care journey—specifically in diagnostics—were mapped out.

71

By approaching IPA as a potential boundary object (Figure 4.9), an artefact was designed that the multiple stakeholders find useful in practice, while it holds different meanings for each stakeholder. Where IPA serves as a translation tool for professionals to better understand the patient, it acts as an exploration tool for the patient to engage in self-reflection on their experience of psychosis, aiming at meaning-making and understanding. At the same time, there is a common understanding of the different stakeholders of IPA as a 'method in the diagnostic process', making it a mediating artefact of perspective-making and perspective-taking.

Figure 4.9. In Pictures Approach (IPA) as a boundary object.

Discussion and Conclusions

Both cases show that the patient perspective can play an essential role in co-design processes in mental healthcare. The patient and professional medical perspectives can coexist by approaching the cases from the viewpoint that potential boundary objects can start perspective-making and perspective-taking processes. From that perspective, a solution can be designed in which both perspectives recognize themselves and experience added value. This view can take co-design, which in mental healthcare practice often does not go beyond a reactive role for the patient, to the next level. While the above cases illustrate this, caution is required, especially in mental healthcare. In both cases, the target group could reflect, describe, and formulate their experiences and needs. The children with ASD were the so-called group of high-functioning children with ASD, and the target group in the case of IPA had already taken significant steps in their recovery journey. In the case of IPA, research is ongoing to confirm if IPA still serves as a mediating artefact when fully integrated into the diagnostic process. The results described in this case reflect the views of the stakeholders and design team in the design workshops, which are significant views but are not the outcomes of research on the effects of IPA as a diagnostic tool in practice. Furthermore, the approach of those studies may not be appropriate for target groups in a crisis. In addition, more and more individuals nowadays believe they know what is good for them based on individual (internet) 'research', which is not always helpful. Boundary theory can help to open an extra perspective when co-designing with mental healthcare target groups. However, careful and well-considered thoughts must always be given to judge which target groups this approach suits. Boundary theory, however, can help balance between artificial co-design and giving shape and meaning to solution directions together with target groups.

Key Takeaways

- Co-design, in which boundary theory is applied, had the potential to open up the patient perspective in mental healthcare design studies and go beyond the reactive participation of end users, as well as to develop mediating interventions that promote perspectives to coexist.
- The added value of boundary theory in co-design lies in the fact that consensus does not need to be reached on perspectives and experiences, which may have an impact on the power imbalances in mental healthcare design studies.
- The two cases demonstrate that individuals with lived experience of a mental health condition can make a significant contribution to design-driven innovation projects and that their insights are essential for developing meaningful concepts.
- Design research practitioners in mental healthcare must remain critical of who this approach is suitable for, as participants in the described cases were relatively advanced in their recovery (case IPA) or highly functioning (case AscapeD). This approach is currently being studied in mental healthcare practice, and the results are still pending.

73

References

• Akkerman, S. F., & Bakker, A. (2011). Boundary crossing and boundary objects. *Review of Educational Research, 81*(1), 132–169. https://doi.org/10.3102/0034654311404435

• Blandford, A., Gibbs, J., Newhouse, N., Perski, O., Singh, A., & Murray, E. (2018). Seven lessons for interdisciplinary research on interactive digital health interventions. *Digital Health, 4*, 2055207618770325. https://doi.org/10.1177/2055207618770325

• Brotherdale, R., Berry, K., Branitsky, A., & Bucci, S. (2024). Co-producing digital mental health interventions: A systematic review. *Digital Health.* https://doi.org/10.1177/20552076241239172

• Carlile, P. R. (2002). A pragmatic view of knowledge and boundaries: Boundary objects in new product development. *Organization Science, 13*(4), 442–455.

• Corstens, D., & Longden, E. (2013). The origins of voices: Links between life history and voice hearing in a survey of 100 cases. *Psychosis, 5*(3), 270–285. https://doi.org/10.1080/17522439.2013.816337

• Cross, N. (2023). Design thinking: What just happened? *Design Studies, 86*, 101187. https://doi.org/10.1016/j.destud.2023.101187

• Fusar-Poli, P., Estradé, A., Stanghellini, G., Venables, J., Onwumere, J., Messas, G., Gilardi, L., ... Maj, M. (2022). The lived experience of psychosis: A bottom-up review co-written by experts by experience and academics. *World Psychiatry, 21*(2), 168–188. https://doi.org/10.1002/wps.20959

• Groeneveld, B., Dekkers, T., Boon, B., & D'Olivo, P. (2018). Challenges for design researchers in healthcare. *Design for Health, 2*(3), 305–326. https://doi.org/10.1080/24735132.2018.1541699

• Huber, M., Knottnerus, J. A., Green, L., Horst, H. V. D., Jadad, A. R., Kromhout, D., ... Smits, H. (2011). How should we define health? *BMJ, 343*, d4163. https://doi.org/10.1136/bmj.d4163

• Islind, A. S., & Lundh Snis, U. (2017). Learning in home care: A digital artifact as a designated boundary object-in-use. *Journal of Workplace Learning, 29*(7/8), 577–587.

- Islind, A. S., Lindroth, T., Lundin, J., & Steineck, G. (2019). Co-designing a digital platform with boundary objects: Bringing together heterogeneous users in healthcare. *Health Technology, 9*(4), 425–438.

- Jensen, S., & Kushniruk, A. (2016). Boundary objects in clinical simulation and design of eHealth. *Health Informatics Journal, 22*(2), 248–264.

- Kajamaa, A. (2011). Boundary breaking in a hospital. *The Learning Organization, 18*(5), 361–377.

- Kuipers, D. A., Terlouw, G., Wartena, B. O., van't Veer, J. T., Prins, J. T., & Pierie, J. P. E. (2017). The role of transfer in designing games and simulations for health: Systematic review. *JMIR Serious Games, 5*(4), e7880. https://doi.org/10.2196/jmir.7880

- Marshall, P., Barbrook, J., Collins, G., Foster, S., Glossop, Z., Inkster, C., Jebb, P., ... Lobban, F. (2024). Designing a library of lived experience for mental health: Integrated realist synthesis and experience-based co-design study in UK mental health services. *BMJ Open, 14*(1), e081188. https://doi.org/10.1136/bmjopen-2023-081188

- Nakarada-Kordic, I., Hayes, N., Reay, S. D., Corbet, C., & Chan, A. (2017). Co-designing for mental health: Creative methods to engage young people experiencing psychosis. *Design for Health, 1*, 229–244. https://doi.org/10.1080/24735132.2017.1386954

- Ritunnano, R., Kleinman, J., Whyte Oshodi, D., Michail, M., Nelson, B., Humpston, C. S., & Broome, M. R. (2022). Subjective experience and meaning of delusions in psychosis: A systematic review and qualitative evidence synthesis. *The Lancet Psychiatry, 9*(6), 458–476. https://doi.org/10.1016/S2215-0366(22)00104-3

- Sajtos, L., Kleinaltenkamp, M., & Harrison, J. (2018). Boundary objects for institutional work across service ecosystems. *Journal of Service Management, 29*(4), 615–640.

- Stompff, G., & Smulders, F. E. (2013, June). The boundary-spanning practice of (user-centered) design. In *Proceedings of the 20th Innovation Product Development Management Conference.*

- Terlouw, G., van't Veer, J. T. B., Kuipers, D. A., & Metselaar, J. (2020). Context analysis, needs assessment, and persona development: Towards a digital game-like intervention for high-functioning children with ASD to train social skills. *Early Child Development and Care, 190*(13), 2050–2065. https://doi.org/10.1080/03004430.2019.164463

- Terlouw, G., Kuipers, D., van't Veer, J., Prins, J. T., & Pierie, J. P. E. (2021). The development of an escape room–based serious game to trigger social interaction and communication between high-functioning children with autism and their peers: Iterative design approach. *JMIR Serious Games, 9*(1), e19765. https://doi.org/10.2196/19765

- Terlouw, G., Kuipers, D., Veldmeijer, L., van't Veer, J., Prins, J., & Pierie, J. P. E. (2022). Boundary objects as dialogical learning accelerators for social change in design for health: Systematic review. *JMIR Human Factors, 9*(1), e31167. https://doi.org/10.2196/31167

- Veldmeijer, L., Terlouw, G., Van Os, J., Van Dijk, O., Van 't Veer, J., & Boonstra, N. (2023). *The involvement of service users and people with lived experience in mental health care Innovation through Design: Systematic review. JMIR Mental Health*, 10, e46590. https://doi.org/10.2196/4659

- Veldmeijer, L., Terlouw, G., van't Veer, J., van Os, J., & Boonstra, N. (2023). Design for mental health: Can design promote human-centered diagnostics? *Design for Health, 7*(1), 5–23. https://doi.org/10.1080/24735132.2023.2171223

- Veldmeijer, L., Terlouw, G., van Os, J., van't Veer, J., & Boonstra, N. (2024). The frequency of design studies targeting people with psychotic symptoms and features in mental health care innovation: A research letter of a secondary data analysis. *JMIR Mental Health.* https://doi.org/10.2196/54202

- Veldmeijer, L., Terlouw, G., van Os, J., Meerman, S. T., van't Veer, J., & Boonstra, N. (2024). From diagnosis to dialogue – Reconsidering the DSM as a conversation piece in mental health care: A hypothesis and theory. *Frontiers in Psychiatry, 15.* https://doi.org/10.3389/fpsyt.2024.1426475

Derek A. Kuipers, Jetse Goris,
Jan-Wessel Hovingh

5. No, I Don't Know Either – New Rituals for Collective Learning: The Design Process as a Temporary Boundary World

DOI: 10.1201/9781003609766

Introduction

Our world is defined by a multitude of interconnected systems and networks. We create, use and have become used to technology to make our lives easier, happier and healthier. These technological interventions have evolved with us into the complex, networked reality that we today call home. Consequently, the challenges associated with change and transformation in this world are also systemic in nature. This means that the causes, consequences and potential solutions to these challenges are in a state of constant interaction. A proposed solution often creates a new set of problems, especially when the deeper systemic consequences of that given solution are not sufficiently taken into account.

It could even be said that today's perceived problems are mostly consequences of yesterday's solutions. Thus, most of these problems are de facto 'designed' problems, and if these deliberate solutions are now creating new problems, it would be safe to assume that we could once again 'design our way out'. After all, *to design*, as Herbert Simon once aptly put it, is *to devise courses of action aimed at transforming existing situations into preferred ones* (Simon, 1996).

But is it reasonable to expect an individual or even a small group of experts to oversee the entirety of systemic dependencies surrounding a perceived challenge and its possible solution? Is it realistic to expect that a desired outcome is attainable through a single project approach, with its intended preferred situation, as Simon said, remaining constant throughout the entire process? It is reasonable to question whether it is possible for anyone to predict and anticipate the reciprocal influences of all systemic variables in advance? This may sound familiar to everyone who has ever been involved in such a process: many solutions are put forward, but it remains unclear what the precise nature of the issue entails. This is where the importance of the focus on learning becomes evident. There is a deeper, collective 'need to know'. If we are conscious of the complexity and contingency here, it is then appropriate to display modesty and admit we may not be that cocksure either. The acceptance of uncertainty is a prerequisite for open learning.

The purpose of this chapter is to direct attention towards the design process with a particular emphasis on the opportunities for learning. Instead of regarding the designer as a universal and directive problem solver and a design process as a necessary path towards a solution, we propose to see a designer as a curator of collective curiosity, who provides optimal preconditions for creative inquiry (Dewey, 2008) and who facilitates a collective reciprocal learning process. Then, from this standpoint, we think the design profession and design processes can be better equipped for dealing with the complex systemic challenges and constraints we see ourselves confronted with today.

Given that design processes are frequently non-linear and entail the consideration of both known and, more significantly, unknown variables and constraints (Schon, 1987), we think it is unavoidable that the process itself entails a learning component. The design process itself reveals a significant amount of information about the given issue, the intended and unintended directions for approaching it and its direct and indirect stakeholders. Instead of a more or less unintended byproduct of a design process, this learning component should be given a prominent place in the design process as a matter of course. Next to 'How do we find a solution?' the question 'What can we learn about the problem?' should be made just as important.

In close collaboration with the Future Design Centre, the Design Driven Innovation Professorship has developed a conceptual framework and process model for the integration of principles of pedagogical practices into design. Over the past few years, this approach has been applied to address numerous real-world issues, resulting in the final iteration of the *Design Driven Innovation Helix*.

Designerly Ways of Learning

Although research-through-design processes, such as design thinking (Cross, 1992), are intrinsically iterative and cyclic, they are focused on finding (prototypal) solutions to a given problem. The notion that designers can acknowledge and embrace uncertainty, sparks the need for learning. Kessels made a significant contribution to this line of thinking by proposing that the design process could be conceptualised as a learning process. The designer gains insight into their own position on the issue, its nature and the appropriate course of action. Moreover, he identifies designing *together* as an 'interesting learning project' for the designer, client, trainers and local managers. This collaborative learning is characterised by the designer directing interactions of a reflexive nature (Poell & Kessels, 2021) .

So, what can be the connection between the design process and the learning process? Designers place great importance on multidisciplinarity, primarily due to the valuable perspectives it brings. However, if facilitating collective learning becomes a designer's objective, it is advisable to adopt a multidisciplinary perspective on the realm of design as well: this is where the design discipline intersects with the educational perspective.

We propose that the role of the designer can evolve from that of a creative problem solver or technical innovator engaged in a design process, to that of a specialised actor and fellow learner within a systemic reality. A designer would no longer involve stakeholders only because it would suit his or her own process of seeking better insights towards preferred outcomes, but instead actively participates with these stakeholders in a learning collective. This involvement of stakeholders in the learning process is paramount, as it provides the preconditions for creating a community that is able to attain shared mental models and shared values. This in turn provides pathways to adopt possible solutions in the long run and provides an environment where possible adverse outcomes can be better negotiated. This learning includes, but is not limited to, the alignment and identification of underlying narratives, linguistic socialisation, and fostering generative emergence.

A Theory of Collaborative Learning

Framing design as a catalyst for collective learning prompts us to delve into the insights offered by the various variants of Cultural Historical Activity Theory (CHAT). This theory is centred on three core ideas: 1) people act collectively, learn by doing, and communicate in and through their actions and artefacts (man-made objects); 2) people make, use and adapt tools of all kinds to learn and communicate; and 3) community is central to the process of making and interpreting meaning, and thus to all forms of learning, communication and action (Vygotsky & Cole, 1978).

The systemic view presented by CHAT (Cole, 1998; Engeström et al., 2016) identifies a number of distinctive components, which significantly increases the number of variables that can create friction, ambiguity, resistance or substantive barriers. These variables are resemblant of the issues causing complexity in the wicked world of designers.

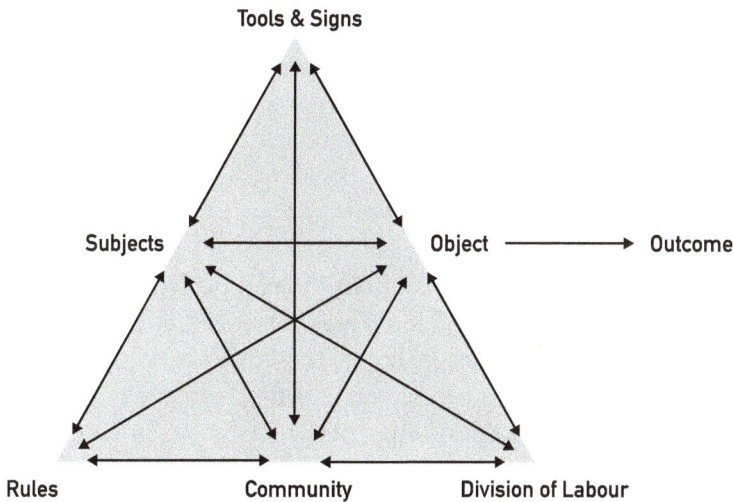

Figure 5.1. Second Generation CHAT (Leontiev).

Building on CHAT, it can be argued that people's deepest beliefs are rooted in the activity system to which they feel they belong. The prevailing narrative in such a system comprises the totality of beliefs, values and frameworks from which a person operates. Crucially, these are often neither explicit nor consciously put on the table. CHAT teaches us that these underlying motives reveal themselves in concrete actions, and often through or by artefacts. In the context of CHAT, the artefact is associated both with the terms *object* and *predmet*. However, they do carry slightly different connotations that are important to understand. The object refers to the broader goal or purpose, while the predmet highlights the specific, concrete focus that engages and motivates individuals within an activity. This constitutes an indispensable link to the designer as sense-maker (Kaptelinin, 2005).

A Designerly Take on Boundary Objects and Boundary Work

As mentioned before, besides the fact that CHAT offers the potential to explore collaborative learning processes (known in CHAT as expansive learning), it also offers explanations for the earlier mentioned challenges when design processes become open, networked and unstable. When the stakeholders in a design process represent different activity systems, incompatibilities in one or more facets may arise. For example, the rules and tools used in one activity system may be considered incompatible, worthless or even hostile by another. Obviously, this does not contribute to a healthy collaborative environment. At the same time though, CHAT also offers pointers that may contribute in overcoming precisely these incompatibilities. Specifically, we distinguish three conceptual approaches for design and the design process in an expansive collective learning process, as proposed by CHAT:

1. Boundary Objects as Intentional Artefacts

A strategy to overcome these friction points in design processes is the application of boundary objects and deploying boundary work activities (Akkerman & Bakker, 2011; Langley et al., 2019). These boundaries can be described *as sociocultural differences that give rise to discontinuities in interaction and action* (Akkerman & Bakker, 2011). Seen in the context of CHAT, a boundary can therefore be described as the demarcation between two or more cultural historical activity systems. Attempts to cross these boundaries result in both challenges to said activity systems as the boundary crossing agents, be they human or non-human: boundary workers or boundary objects. However, boundaries can be crossed on the object-level (Star & Griesemer, 1989), even though these objects may not be primarily intended to be used in that way.

This implies that objects intended as boundary objects must have multistable qualities (Ihde, 2016) in order to address the unique values of said activity systems. Such an object will have a use and a meaning aligning with the tools and rules in one activity system, while it has a different use and meaning in another, all while supporting a shared objective (predmet) for both activity systems. This means that, if the shared objective would be collective learning, all objects and activities should be chosen, designed and deployed with this collective learning in mind.

2. Collaborative Boundary Work for Intentional Learning Activities

Collaborative boundary work refers to the process in which individuals or groups from different domains, disciplines, or sectors actively engage in defining, negotiating, and managing boundaries to work together effectively. This involves bridging differences in knowledge, values, practices, and goals to foster cooperation and achieve shared objectives.

When describing boundary work as a strategy for multidisciplinary collaboration, Langley indicates collaborative boundary work as practices through which groups, occupations, and organizations work at boundaries to develop and sustain patterns of collaboration and coordination in settings where groups cannot achieve collective goals alone (Langley et al., 2019). Langley explicitly names *learning* as one of the consequences of collaborative boundary work.

Here the actual activities can be a collaborative design activity that facilitates the learning process. The new role of the designer here is that of learning architect and facilitator of the group learning process.

3. Temporary Boundary Worlds

The present argument posits that the act of convening across boundaries can be enhanced through the undertaking of collaborative design activities situated within a *temporary boundary world*. To this end, we propose that such encounters should be structured around a series of design activities focused on collaborative learning, operating in a transient and conceptualised 'boundary world', which is not intended to be an arena of conflict resolution, but one in which the aim is to gain shared insight and knowledge for all participants involved.

Instead of seeking just the communicative connection (Star & Griesemer, 1989) on the (boundary) object level, or developing and sustaining patterns of collaboration, we propose to treat such a design process as a cultural activity system in itself: a temporary community, where stakeholders and participants work towards a shared objective using appropriate, shared tools and shared rules.

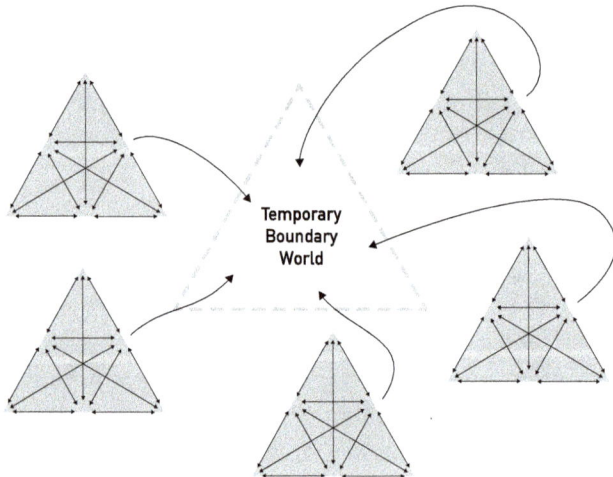

Temporary
Boundary
World

Figure 5.2. The temporary boundary world, situated amidst multiple cultural activity systems.

This creates both a necessity for finding a deeper appreciation of each other's narratives as it facilitates the acquisition of mutual understanding through the examination of each other's perspectives, values, rituals and motivations. In the context of design activities, this subject matter is revealed with greater rapidity and clarity.

The Design Helix

In recent years, the focus of our research group has been on developing a visualisation of the collective learning process as it occurs within the context of a design process. It is our contention that the Helix contributes to the broader design discourse by focusing on expansive learning, driven by design activity. Based on the principles of the PDCA-cycle (Shewhart 1980), the Design Research Helix is a conceptual framework that integrates design and learning.

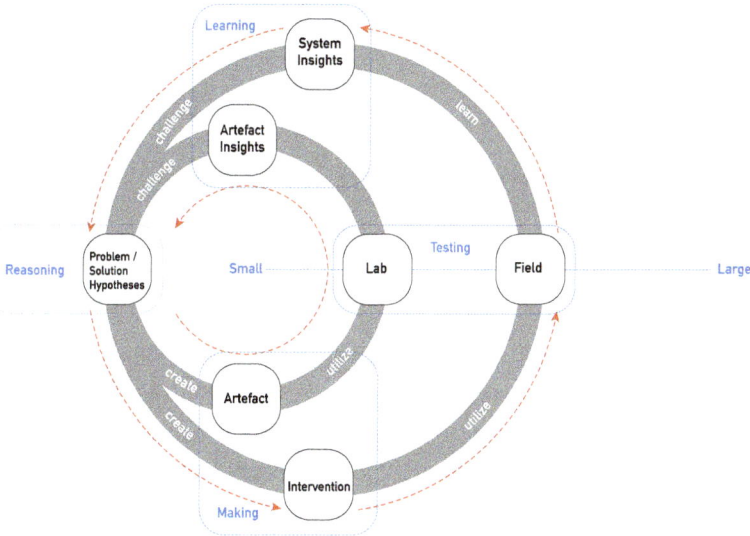

Figure 5.3. The Design Driven Innovation Helix for design and learning.

The model comprises a number of focus areas, which are interconnected and mutually reinforcing. It is important to note at the outset that there is no distinct beginning or endpoint. Each node may be regarded as an entry point for the design-learning cycles.

We shall commence with the lower half. In this, the artefact is regarded as the design-in-the-small, deliberately loaded and designed as a boundary object for deployment in order to facilitate learning from and with. The subtlety here is that this artefact can also be understood as part of a design activity (design-in-the-large). By 'intervention', we refer to the artefact as part of a design research activity, encompassing the entirety of the intent of the design activity (Klabbers, 2003).

It is evident that a rationale or starting point is a prerequisite for the selection of an artefact or intervention. This is derived from the design/problem hypothesis, which represents an initial, non-fixed perspective on the matter at hand (Stompff et al., 2022). The solution part of this hypothesis can be seen as the frame (Dorst, 2015): it comprises a rationale of the desired situation and how this situation could be achieved. It is crucial to recognise that this version of the frame is, at best, an educated guess, as this fosters a learning-oriented attitude. Furthermore, it is crucial to recognise that this hypothesis, which posits a problem and a solution, has yet to be informed by the insights informed by other, as yet unknown, activity systems.

The deployment of the artifact or intervention is conducted through a process of testing. If the focus is on the tuning and operation of the artefact, this takes place in the 'lab'. However, this should not be understood as a physical place, but rather to a smaller setting where the preparation of the intervention is investigated. By 'testing' the 'intervention' in the 'field', we refer to the deployment of the intervention in a broader context of the then involved stakeholders. Here, data collection is done by recording the generated meaning-making processes.

The top of the helix shows where learning takes place. Again, it is important to note that interpreting the data and results is not a one-way activity for the design researcher: interpreting, mapping and visualising the data is a broader activity that should also involve stakeholders. Context is key. Again, learning can focus on design-in-the-small, aligning the learning outcomes of the artefact with the insights gained, or design-in-the-large: the insights gained from the 'intervention'.

In accordance with the established iterative nature of the design process, the insights gained are employed to challenge the initially proposed frame. These insights offer a novel perspective on the fundamental challenge at hand. This represents a pivotal aspect of design learning, whereby the incorporation of new information may prompt a shift in the problem/solution hypothesis, potentially leading to the involvement of additional stakeholders, a revised perspective on the problem, or the introduction of an alternative frame.

The Fjouwerhoeke Case

In order to demonstrate how the design process can act as a catalyst for learning, we will examine the successful case of Fjouwerhoeke. Fjouwerhoeke is a constituent member of a larger primary school trust, known as Arlanta. The specific school is situated in a rural area, characterised by a declining population and a reduction in the availability of services and amenities. The chair of the executive board posed the following initial question:

Given the availability of funding to construct a new school in a rural region with a declining population, how might the school differentiate itself and maintain a sufficient pupil population in the future?

In rural areas, elementary schools are a long-standing feature of the village landscape. Each village typically possesses its own educational institution, which is widely regarded as the social and cultural centre of the community. The profound emotional connections that these schools engender within the village community foster a resilient yet internally focused cultural identity,

89

where the closure of a school is often seen as a symbol of the village's decline. The process of merging with other nearby schools is frequently a challenging and emotionally laden task, a transition that must be managed with great care.

Learning began where the different ideas extended far beyond that of the physical building alone. The school, or rather, thinking about the new school, functioned as a *boundary object* because involved parties had their own view of what a school is, or should be. Some ideas were deep-rooted, conflicting and seemingly incompatible: it quickly became clear that before we could talk about the building, we needed to learn about everyone's perspectives. In a series of design activities, we searched with stakeholders for a suitable frame for the new Fjouwerhoeke that did justice to all views. It is important to note that the willingness to participate and the degree of commitment increased when the focus shifted from the physical solution (a new building) to the exploration of what a new school would mean for the community.

Building on the mutual trust and shared perspective, *collaborative boundary work* led to the actual design of the expressions appropriate to the developed frame. The new frame presented the team with new insights and new potential partners. In collaboration with the parents of the pupils, the team identified four new partners: the library, the local childcare organisation, the agricultural vocational school and the local community of the four surrounding villages. The introduction of new partners signifies the crossing of boundaries between distinct activity systems. The initial design sessions were conducted with the objective of investigating the potential for synergy between the envisaged partners. A member of each organisation's executive board convened at the location of one of the partners. The potential added value and synergy between partners was explored through the completion of a comprehensive matrix, which mapped all possible combinations of partners. Following the ideation of various combinations, a set of *synergy combinations* was selected for utilisation in the subsequent session. Rather than conceptualising this design intervention as a mere crossing of boundaries, it could be perceived as a method of gradually acclimating the board members to the concept by facilitating interactions at the boundaries between their respective organisations.

With an explored environment, unforeseen collaborations and partners, and a supported perspective, the collective learning continued through the design of potential future scenarios in a *temporary boundary world*. The design task required the participants to imagine a scenario in which the new Fjouwerhoeke is a resounding success five years into the future. They were then asked to write a script for a podcast in that future, in which a reporter visits the school for a day and experiences the new school. The participants were instructed to include concrete examples of the synergy between the various partners involved in the construction of the new school building. In order to provide guidance and inspiration, the designer created an initial, provisional draft of the script, which was then made available for amendment by the partners. The final draft was completed and subsequently reviewed, finalised and recorded with the participation of all actual stakeholders. The result was an illustrated podcast on YouTube, created by the collective of people and envisioned partners of the new Fjouwerhoeke school.

The collective act of scriptwriting thus engenders the possibility of a new future. In the course of this act, the partners collectively explored the interests, rules, community and division of labour of the future collaboration. The final outcome was presented to the municipal council of North East Fryslân during an information meeting, during which the details were discussed in depth. The meeting was meticulously structured to elucidate the collective intention and objective of the diverse partners.

Future Work

Furthermore, Sanne Akkerman states the following (translated from Dutch):

"System as structure: to what extent does thinking in systems apply to more temporary and creative practices of action, where several (independent) actors are active, and where the object they focus on also changes (as in the case of the open source movement on the internet)? (...) Engeström has already suggested that a fourth step in CHAT's development may be needed. This indicates that CHAT itself is also a cultural-historical theory in permanent development."

(Redactie Boom Management, 2020)

It is notable that there are clear similarities between our observations, suggestions, experiences and those made by Akkerman. This may indicate the potential for further development of CHAT, with a particular focus on creative learning processes.

This combination of theoretical perspectives has significant potential: a collective learning process that employs design as a mechanism to generate knowledge has the potential to facilitate a transformation in our cultural narratives. It may encourage the admission that present-day transitions and challenges are often so complex and multifaceted that immediate answers are not feasible, but that there is a common interest in learning how to bring them about. It may also engender a system-wide curiosity that gives rise to a collective aspiration to learn instead of a narrow focus on an individual solution.

Acknowledgements

The Design Driven Innovation Helix framework was developed over several years through a process of collaboration, discussion and refinement with a number of esteemed colleagues at the Future Design Center of NHL Stenden University of Applied Sciences. We would like to acknowledge the contributions of Steven de Rooij and Boudewijn Dijkstra in particular.

Key Takeaways

It should be noted that this chapter has addressed a considerable number of topics, at times advancing significantly and leaving certain subjects underexplored. The primary objective has been to illustrate that the design process can be conceptualised as a learning journey, which presents a multitude of possibilities and challenges, each of these warrants further investigation and detailed analysis.

In accordance with the designation of this chapter, the following principles are thus established:

- It is incumbent upon the designer to acknowledge that a singular solution is not a static entity, but rather a dynamic entity that evolves through a process of learning.
- The designer is a member of a collective engaged in a process of learning, occupying a position of facilitation and inquiry within this collective.
- The objective of this learning process is to identify a supported solution-path, as opposed to a definitive solution.
- Joint learning occurs within a secure setting, a temporary boundary world, where delegates from their respective activity systems engage in collaborative learning with and from one another in a tailored design endeavour. To effectively guide and delineate the nature and trajectory of design activities, new rituals are essential to provide meaningful and succinct direction to the proposed learning process.

In framing the design process as a learning activity, new tooling, methodology and theoretical frameworks become available to designers: didactic, psychological and sociological fields of expertise such as Social-Constructivism, Cultural Historical Activity Theory and Boundary Work. The positioning of the design process in a temporary 'free place', or σχολή in Greek (school), enables a shift in focus from solution-based thinking to process-based transformation. The design process itself can be considered a carrier of change through collective learning.

References

- Akkerman, S. F., & Bakker, A. (2011). Boundary crossing and boundary objects. *Review of Educational Research, 81*(2), 132–169. https://doi.org/10.3102/0034654311404435

- Cole, M. (1998). *Cultural psychology: A once and future discipline.* Harvard University Press.

- Cross, N. (1992). *Research in design thinking: Proceedings of a workshop meeting held at the Faculty of Industrial Design Engineering, Delft University of Technology, the Netherlands, May 29–31, 1991.*

- Dewey, J. (2008). *The later works of John Dewey, Volume 12, 1925-1953: 1938, Logic: The theory of inquiry.* SIU Press.

- Dorst, K. (2015). *Frame innovation: Create new thinking by design.* MIT Press.

- Engeström, Y., Lompscher, J., & Rückriem, G. (2016). *Putting activity theory to work: Contributions from developmental work research.* Lehmanns Media.

- Ihde, D. (2016). *Husserl's missing technologies.* Fordham University Press.

- Kaptelinin, V. (2005). The object of activity: Making sense of the sense-maker. *Mind, Culture, and Activity, 12*(1), 4–18.

- Klabbers, J. H. G. (2003). Gaming and simulation: Principles of a science of design. *Simulation & Gaming, 34*(4), 569–591.

- Langley, A., Lindberg, K., Mørk, B. E., Nicolini, D., Raviola, E., & Walter, L. (2019). Boundary work among groups, occupations, and organizations: From cartography to process. *Annals, 13*(2), 704–736.

- Redactie Boom Management. (2020, August 31). De Activity theory van Yrjö Engeström (longread). *Boom Management.*https://boommanagement.nl/artikel/de-activity-theory-van-yrjo-engestrom-longread/

- Poell, R., & Kessels, J. (2021). *Handboek human resource development: Organiseren van het leren.* Lannoo Meulenhoff - Belgium.

- Schön, D. A. (1987). *Educating the reflective practitioner: Toward a new design for teaching and learning in the professions.* Jossey-Bass.

- Simon, H. A. (1996). *The sciences of the artificial.* MIT Press.
- Star, S. L., & Griesemer, J. R. (1989). Institutional ecology, 'translations,' and boundary objects: Amateurs and professionals in Berkeley's Museum of Vertebrate Zoology, 1907–39. *Social Studies of Science, 19*(3), 387–420.
- Stompff, G., van Bruinessen, T., & Smulders, F. (2022). The generative dance of design inquiry: Exploring Dewey's pragmatism for design research. *Design Studies, 83*, 101136.
- Vygotsky, L. S., & Cole, M. (1978). *Mind in society: Development of higher psychological processes.* Harvard University Press.

Remko van der Lugt, Lenny van
Onselen, Christa van Gessel

6. Impact by Toolkits as Generative Knowledge Carriers in Healthcare

DOI: 10.1201/9781003609766

Introduction

In design research grant proposals, researchers often mention that they will develop toolkits to transfer the insights gained. The motivation for proposing toolkits is that design professionals like to use practical and actionable tools in co-design projects, for example, card sets with design methods (Roy & Warren, 2019) and design games (Brandt, 2006). They aim to bridge the research-practice gap (Smeenk et al., 2024).

In an earlier contribution (van der Lugt, 2022), we discussed how design facilitation involves developing materials that allow participants to engage in the co-design activities in a designerly manner, in addition to regular process facilitation (Aguirre et al., 2017)). Similarly, designly ways to stimulate the impact of the results of design research projects in practice, crystallize insights into materials that invite further creative application and stimulate making own adaptations. Research insights can be made accessible and actionable though descriptive and depictive materials (e.g. card sets), but also through materials that transfer insights by making them actionable in their use (e.g. socionas, sets of figurines to explore hidden dynamics (van Gessel et al., 2018). Collections of such materials are referred to as toolkits (Sanders & Stappers, 2014).

We regard such toolkits as 'generative knowledge carriers': knowledge carriers because they are intended to transfer insights from researchers to co-designers and/or practitioners; and generative, because they are intended to feed the creative process of its users. Toolkits aim to make insights from research available in an open-ended way, so they can easily be appropriated and re-organized to suit the practice and question at hand.

In this chapter, we discuss the qualities and design considerations of toolkits as generative knowledge carriers for healthcare professionals. First we provide some theoretical considerations on toolkits as generative knowledge carriers. Then we present and discuss two toolkits that were developed in co-design research projects to impact professional healthcare practice. For each toolkit, we discuss the design considerations for their functioning as generative knowledge carriers. Finally, we will provide some overarching insights regarding the design and functioning of toolkits for professional practice.

Theoretical Background

Toolkits in Design

Within the field of design research and innovation, research on toolkits has focused on its use in the co-design process, by providing participants (future users) with materials to investigate and express their experiences (Mattelmakki, 2006; Stappers & Sanders, 2012), or to enable them to create prototypes of a new product or service, to inform the professional design innovation process as lead users (e.g. Von Hippel, 1986). Within the field of HCI, various publications address the use of toolkits in online environments to 'design your own product' as part of the mass-customization movement (e.g. Franke & Piller, 2004), which allows users to 'design' their personalized product configuration based on a variation of components.

The toolkits we discuss here attempt to inform, inspire or change professional practice of non-designers. These toolkits attempt to transfer insights from design research to improve the professional practice of healthcare professionals. They introduce designerly ways of working into the practice of healthcare professionals.

Toolkits as Intermediate-Level Knowledge Carriers

The knowledge generated in design research is often situated in between the generality of a theory and the specificity of a singular specific situation. Hook and Lowgren (2019) name this 'intermediate-level knowledge', and mention that this kind of knowledge is particularly valuable in generative design activity. Zielhuis (2023) identified three ways in which design research collaborations contribute to design practice: by taking part in the design research project; by interacting with research partners, for instance, by means of training or workshops; or by interaction with 'knowledge products'. The toolkits we consider here fit in

99

this third category. However, the categories are not exclusive: For instance, Brandt (2006) found that toolkits become more effective for practice when users and experts are involved in the process of designing them (Brandt, 2006).

The gap between research and practice is typically referred to in terms of abstraction—as in, practitioners need fewer abstract concepts to utilize in their practices, or in terms of communication —as in, researchers and practitioners speak different languages. Smeenk et al. (2024) critically review this research-practice gap and provide two additional characteristics: alignment of needs: the kinds of knowledge needed shift throughout the design process—as in, researchers intend to describe or explain the world, while designers need generative knowledge that inspires new concepts. Finally, they identify the characteristic of supporting local knowledge production—as in, it is not solely about utilizing the produced knowledge, but utilization also involves situated knowledge production. These characteristics can help guide our understanding of the function of toolkits for professional practice.

We regard toolkits as 'knowledge carriers'. We observed that toolkits integrate and disseminate findings from research and design into professional practice, by explanations of the knowledge in text and graphics, or by inviting new practices by means of physical tools. The materials in these tools function as knowledge carriers in their own right as Buur et al. (e.g. 2022) elucidate in their research on 'theory instruments': the materials afford use that transfers insights from the resource without having to explain or depict the insight itself. This is much in line with the famous quote attributed to R. Buckminster Fuller: *"If you want to teach people a new way of thinking, don't bother trying to teach them. Instead, give them a tool, the use of which will lead to new ways of thinking."*

Toolkits as Generative Knowledge Carriers

We see such toolkits as 'generative' in two ways. First, because they are intended to feed the creative process of its users. Toolkits aim to make insights from research available in an open-ended way, so they can easily be appropriated and re-organized to suit the question at hand. This way, they can bring generative components and potential for novel connections for the professionals who want to innovate their professional practice (Sleeswijk Visser et al., 2007).

Second, 'generative' can refer to the utilization of creative, design-like, activities to elicit deeper insights from users, in the context of the healthcare professional context, often clients or patients. This connects to the generative techniques approach to have people 'make' something, then talk about it, which allows to make explicit tacit knowledge and latent needs. Such tools are also generative because they provide empathy, information and inspiration to the design process (Sleeswijk Visser et al., 2005). The question remains, to what extent can practitioners stretch their practice into the use of such open-ended, creative tools?

Cases

Below we will reflect on the qualities as generative knowledge carriers of two toolkits developed for professional practice in children's physical therapy. The toolkits that we include as cases aim to transfer insights from design research to professional practice. We limit ourselves to physical toolkits developed in co-creation with practitioners in projects by the Co-Design Research Group of HU University of Applied Sciences Utrecht. This gives us first-hand insight into both the research, design and development of the toolkits, as well as their application in the design process and practice. A second criterion for inclusion is that the toolkits have been reported in a professional or academic publication, which allows readers to gain further background information.

101

Case 1: What Moves You?! Stimulating Children with Disabilities to Play in Their Natural Living Environment

The What Moves You toolkit was developed to fill the need for paediatric physiotherapists (PPTs) to move their practice of supporting children with disabilities beyond the PPT practice space. Children with disabilities in general move too little, because of a variety of reasons, yet movement therapy for children is largely based on consults in the practice space. The design brief was to develop a toolkit to support PPTs and community sports coaches to stimulate movement in the daily living environment. The project was initiated as a follow-up project of a PhD project on the physical fitness and physical behavior in wheelchair-using youth (Bloemen, 2017).

The design research project was executed with a core team of healthcare researchers and co-design researchers and -professionals. In a larger co-design group of 25 people, representatives from children's therapists, community sports coaches, and parents were involved. This group came together every four months for shared synthesis and decision-making. A case study description of the process was reported in a rehabilitation science journal (Bolster et al., 2021).

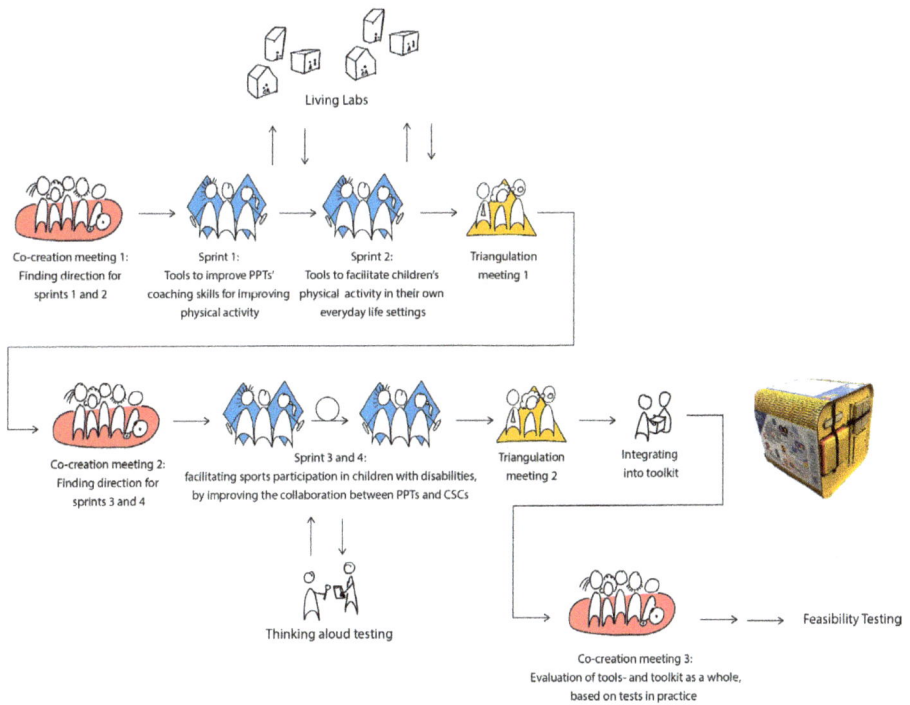

Figure 6.1. The co-design research process in the What Moves You? Project (Bolster et al., 2021).

The project led to, quite literary, a toolbox: a large cube-shaped box, because we wanted the toolkit to be obviously present and at hand in the PPT practice space) with eight physical tools, accompanied by online videoclips and an app for seeking contact between PPTs and community sports coaches. In addition, the toolkit consists of a set of 'knowledge cards' that contain insights from research, with connections to theory, quotes from the qualitative inquiry, and connections with the tools.

The tools intend to provide practical workforms to support (talking about) the practical therapeutic work of the PPT. For instance, sets of green, yellow and red referee cards, to hand out to the children in order to give feedback to people that they are able to do things by themselves and do not need to be 'saved' (for instance getting on the bus).

103

The toolkit prototype was used for feasibility testing, and then further developed into an off-the-shelf kit that can be ordered from a therapy materials company (https://www.alprovi.nl/therapie/spel-en-bouwmaterialen/toolbox-wat-beweegt-jou) (see Figure 6.2).

Figure 6.2. Left, the protoype toolkit: All the tools embedded in a wooden box with instructions on the outside. Right: The final product: a more traditional cardboard box with various tools inside.

The toolkit is intended to provide a set of tools to be applied in operational practice, usually during client consultation. Even though a light structure to guide that process, they provide flexibility regarding how and when to apply them. This is intended to support the PPTs in creatively adapting tools into their practice. The knowledge cards provide intermediate level knowledge that both connect to particular instances and experiences (by means of e.g. quotes and photographs) and to theory (by means of references). Some tools have the qualities of generative techniques to support uncovering tacit knowledge during consultation. For instance, a board with materials on which children can depict their daily environment and their day journey through this environment, thus providing the PPT with insight for where to direct an intervention (see Figure 6.3).

Figure 6.3. Prototype of a 'conversation placemat' as part of the What Moves YOU?! Toolkit: a generative technique intended for the PPT's professional practice.

PEBBLES: ParEntal Beliefs About Their Baby, Lifestyle and Experience Study

In the PEBBLES project, we had the opportunity to study the impact of co-designing a toolkit on the participants. In PEBBLES, a toolkit was designed together with paediatric physiotherapists (PPTs) from six practices, PPTs researchers, knowledge partners, parents, and relevant stakeholders (e.g. Dutch Association for PPTs). In the PEBBLES project, parents, PPTs, and co-designers co-created knowledge and tools to support PPTs and parents in discussing thoughts, ideas, and actions of parents concerning motor development: the so-called parental beliefs and practices. This can be difficult, as parental beliefs are primarily implicit, and PPTs may not see it as part of their role and/or they do not feel competent to address beliefs during consults (Boonzaaijer et al., 2024).

105

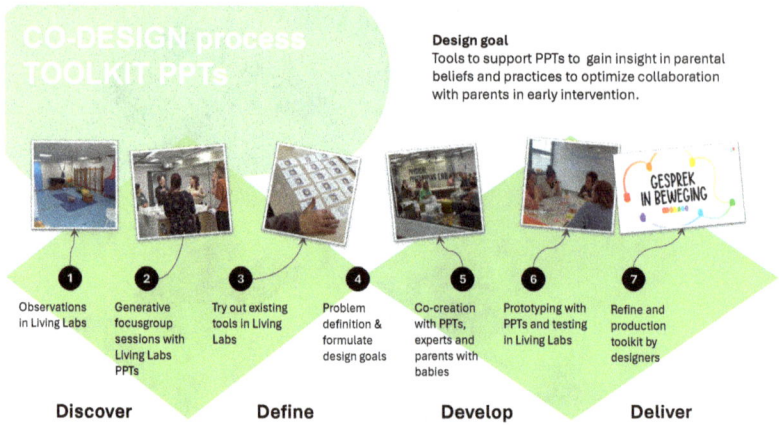

Figure 6.4. The PEBBLES Co-Design research process.

In generative focus groups and interviews with parents and PPTs, insights were collected to analyse the problem (see Figure 6.4). The six PPT practices experimented in context with existing tools (e.g. value cards, questionnaire) which resulted in design criteria for new tools. After this, four co-creation sessions with short iterations were carried out to develop new tools, which were applied and tested by a larger group of PPT's in Living Labs (Higgins & Klein, 2011).

Insights from research from PEBBLES and co-design methods were used as input to co-design the toolkit with the PPTs from the living labs. Also, their experiences as PPT practitioners were used as input for designing the toolkit. In three iterations, prototypes of tools were designed, tested in the living labs, and feedback was used to develop the tools into the toolkit Conversation in Motion (see Figure 6.5). The insights from the tests were also used to further develop the toolkit.

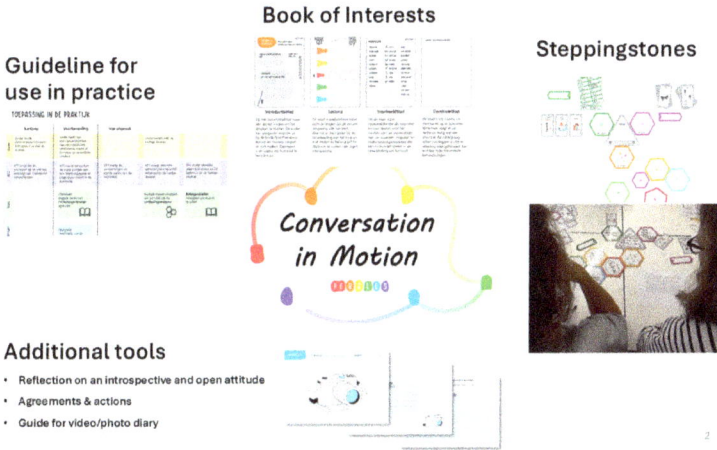

Figure 6.5. The PEBBLES toolkit called Conversation in Motion (Gesprek in Beweging) includes three main tools, a training for the toolkit, and additional tools.

The toolkit Conversation in Motion includes a guideline for use in clinical practice, where you can see in what phase of your intervention you can use the tool according to the purpose. Also, a Booklet of Interests was made, where you can give parents the booklet after the first time you meet and ask them to look at the booklet and fill out the questions in the booklet for the next time. This ensures that parents can look at it and discuss it at home and have the time to think about it. The Steppingstones is a conversation tool to make a map of the beliefs and concerns of parents. This is an open-ended tool that PPTs can use differently per situation to suit the questions at hand. The PPTs can decide to use different cards (Central Question, Parental beliefs, Parental Practices and/or Social Context) to deepen the beliefs of parents on the motoric development of their child or they can map out the characteristics of the case. Additionally, a training program helps PPTs to adopt the knowledge and toolkit for their practice.

The tools stimulate PPTs to explore the children's developmental context and parental beliefs. The tools are generative as they stimulate empathy and provide information that is essential to create an adequate treatment plan for the patient AND their parents. The Booklet of Interests provides information to the PPTs about the beliefs and concerns of parent(s) and other caretakers of the child. In sum, the Conversation in Motion stimulates PPTs to be more explorative and creative in setting up the development plan for their patients. Co-design methods (e.g. service blueprint, socionas, value cards and cultural probes) were used as inspiration for the tools, therefore, the PPTs unconsciously become familiar with these co-design methods. We assume that the PPTs could later apply the explorative and creative approaches in their other PPT practices. Our hopes are that PPTs who participated in the co-creation of the toolkit would use the co-design approach when they see new challenges in their work.

Discussion of Insights

Generative Design Tools Can Bleed into Professional Healthcare Practice

We find that oftentimes generative process tools, intended and used as part of the co-design process, find their way into toolkits intended for practice. For instance, in the What Moves You?! Toolkit, materials are included to describe the daily living environments and the obstacles to moving freely that are encountered, allowing for a generative conversation (see Figures 6.3 and 6.5). We had used the tool as part of the design research process, in order to gain an understanding of the daily experiences and context of a variety of children with disabilities. But the PPTs were so enthusiastic about the tool that they wanted to use it to support conversations in their own practice. In this way, the tools involved bleed downwards in Stappers' meta-design framework (Stappers & Hofman, 2009) from the practice of the designer to the practitioner.

We also saw this in another project with speech therapists (Singer et al., 2022). One of the first co-design activities involved a context-mapping session with the speech therapists, in which they created some sort of personal value map. Rather than as a tool to share their experiences in the co-design process, they saw this as a prototype of the final tool. In any case, these comments strengthened our understanding that the generative ways of working fit the children's therapists professional environment very well.

The Process of Making the Toolkit Can Make at Least as Much Impact on Professional Practice as the Resulting Toolkit Itself

The project objective of developing a toolkit for practice creates a goal in the conceptual realm that allows to bring together thinking about theory and practice, both integrating existing knowledge and experiences, and using that understanding to create concepts. As such, the *doing* of the design of the tools and toolkits provides learning about how the practitioners want their practice to be. And this can have an impact beyond the application of the tools in practice. Godfroij et al. (2024) observed that professionals involved in co-design projects generally develop design ability, and that the research setting (e.g. Living Lab, or Community of Development) contributes to this development. Indeed, the PEBBLES project reported an increase in design ability by practitioners by means of being engaged in the development process. The collaboration in a research setting was experienced as a safe environment in which inter-vision with healthcare colleagues could take place, and in which shared reflection provided insights into personal actions. Besides, the healthcare professionals explicated that they gradually became more aware of their own contributions in the collaboration with co-design researchers. As a PPT put it: *"I am not sure what you want me to do, but I simply can offer practical knowledge. I like the sessions! It is something to which I can contribute".*

Overall, we observed that all participants in the PEBBLES project design ability, either through research insights, participating in co-creation sessions and/or facilitating the co-creation sessions and involving parents as experts of experience. Thus, not only will the toolkit itself impact professionals to change their practices, but in co-creating the toolkit, the PPT researchers and the PPT from the Living Labs were affected in their design ability. The enhanced design ability enables practitioners to reshape their practice.

109

Toolkits Are Not Always Used as Intended, and that Is OK

We find that practitioners can get kind of greedy when they see a nicely developed toolkit. They want to own it, to have it available in their practices. Some toolkits indeed have found their way into professional practice and are used regularly. Other toolkits find their way to the practice's bookshelves and are mostly visited for reference and inspiration, rather than in the direct operations of the practice. The latter is not necessarily a failure of impact: the impact may just be different. Toolkits can be a strong intermediate knowledge means to bring insights from design and research to professional practice. A book or scientific article provides knowledge in a linear manner, while toolkits provide a mix of materials that allow for cognitive scaffolding (van Dijk & van der Lugt, 2013) in a more dynamic manner. Toolkits often contain tools with specific intentions of use, instructions, theoretical grounding and, importantly, examples of their application. This allows practitioners to bridge the gap between practice and theory by dwelling within the conceptual context (van Beest et al., 2022), rather than being forced into either practice or theory contexts.

Healthcare Professionals Don't Mind Using Design Researchers' Toothbrushes', but It Helps If They Can Make Them Their Own

A critical note on the proliferation of toolkits is the—perhaps overused—proverb that "tools are like toothbrushes: every designer has them, but nobody wants to use someone else's". Design researchers like to create toolkits, but ready-made toolkits are not necessarily what the design practitioners want. However, in our experience, this toothbrush issue doesn't necessarily apply to non-design practitioners, who are often more than happy to adopt the toolkits. Still, in several projects we nowadays emphasize the possibility of professionals making toolkits their own, by adding their own contextualized knowledge and tools, which connects to the abovementioned aspect of local knowledge creation (Smeenk et al., 2024).

110

Conclusion

In this chapter, we reflected on toolkits that we developed as deliverables in applied design research projects for healthcare. These toolkits impact practice, by introducing a generative mindset into the healthcare professionals' practices, by making insights available by being 'at hand', and by both enhancing the domain knowledge and the design ability of the domain professionals. However, further research will be needed to investigate how toolkits from design research affect professional practice over a longer time period.

Key Takeaways

The key take-aways on the use of toolkits as generative knowledge carriers are explained in more detail in the section above:

- Generative design tools can bleed into professional healthcare practice.
- The process of making the toolkit can make at least as much impact on professional practice as the resulting toolkit itself
- Toolkits are not always used as intended, and that is OK
- Healthcare professionals don't mind using design researchers' 'toothbrushes', but it helps if they can make them their own.

References

- Aguirre, M., Agudelo, N., & Romm, J. (2017). Design facilitation as emerging practice: Analyzing how designers support multi-stakeholder co-creation. *She Ji: The Journal of Design, Economics, and Innovation, 3*(3), 198–209.

- Bloemen, M. A. T. (2017). *Physical fitness and physical behavior in (wheelchair-using) youth with spina bifida* (Doctoral dissertation, Utrecht University).

- Bolster, E. A., Gessel, C. V., Welten, M., Hermsen, S., Lugt, R. V. D., Kotte, E., ... & Bloemen, M. A. (2021). Using a co-design approach to create tools to facilitate physical activity in children with physical disabilities. *Frontiers in Rehabilitation Sciences, 2*, 707612.

- Boonzaaijer, M., Suir, I., Bolster, E., van Onselen, L., van der Schoot, L., & Godfroij, B. (2024). PEBBLES: ParEntal Beliefs about their Baby, Lifestyle and Experience Study: A study protocol. Presented at *European Academy of Childhood-Onset Disability Annual Meeting, 29 May–1 June 2024.* Available at https://eacd2024.org/scientific-programme/
- Brandt, E. (2006). Designing exploratory design games: A framework for participation in participatory design? In *Proceedings of the Ninth Conference on Participatory Design: Expanding Boundaries in Design - Volume 1 (PDC '06)*(pp. 57–66). Association for Computing Machinery.
- Buur, J., Kjærsgaard, M., Sorenson, J. E., & Ağça, A. Ö. (2022). Studying interaction design practices with theory instruments. In *Design Research Society 2022.* Design Research Society.
- Franke, N., & Piller, F. (2004). Value creation by toolkits for user innovation and design: The case of the watch market. *Journal of Product Innovation Management, 21*(6), 401–415.
- Godfroij, B., van Onselen, L., & Cremers, A. (2024). Developing design ability for innovation in healthcare practices. In *Proceedings of Relating Systems and Design (RSD13): Rivers of Conversation, October 2024.*
- Higgins, A., & Klein, S. (2011). Introduction to the living lab approach. In Y. H. Tan, N. Björn-Andersen, S. Klein, & B. Rukanova (Eds.), *Accelerating global supply chains with IT-innovation* (pp. 1–15). Springer.
- Mattelmäki, T. (2006). *Design probes.* Aalto University.
- Roy, R., & Warren, J. P. (2019). Card-based design tools: A review and analysis of 155 card decks for designers and designing. *Design Studies, 63*, 125–154.
- Sleeswijk Visser, F., Stappers, P. J., Lugt, R. V. D., & Sanders, E. B. N. (2005). Contextmapping: Experience from practice. *CoDesign, 1*(2), 119–149.
- Sleeswijk Visser, F., van der Lugt, R., & Stappers, P. J. (2007). Sharing user experiences in the product innovation process: Participatory design needs participatory communication. *Creativity and Innovation Management, 16*(1), 35–45.

112

- Sanders, E. B. N., & Stappers, P. J. (2012). *Convivial toolbox: Generative research for the front end of design.* BIS.
- Singer, I., Klatte, I. S., de Vries, R., van der Lugt, R., & Gerrits, E. (2022). Using co-design to develop a tool for shared goal-setting with parents in speech and language therapy. *International Journal of Language & Communication Disorders, 57*(6), 1281–1303.
- Stappers, P. J., & Hoffman, R. R. (2009). Once more, into the soup. *IEEE Intelligent Systems, 24*(5), 9–13.
- van Beest, W., Boon, W. P., Andriessen, D., Pol, H., van der Veen, G., & Moors, E. H. (2022). A research pathway model for evaluating the implementation of practice-based research: The case of self-management health innovations. *Research Evaluation, 31*(1), 24–48.
- Van Dijk, J., & van der Lugt, R. (2013). Scaffolds for design communication: Research through design of shared understanding in design meetings. *Artificial Intelligence for Engineering Design, Analysis and Manufacturing, 27*(1), 1–15. https://doi.org/10.1017/S0890060413000024
- Van der Lugt, R. (2022). The designer as a facilitator of change. In J. P. Joore, G. Stompff, & J. Van den Eijnde (Eds.), *Applied design research: A mosaic of 22 examples and interpretations focussing on bridging the gap between practice and academics* (pp. 1–15). CRC Press. https://doi.org/10.1201/9781003265924
- Van Gessel, C., de Vries, R., & van der Lugt, R. (2018). Sociona's, bringing the systemic view into the design for health. In *Proceedings of RSD7 Conference "Relating Systems Thinking and Design"*, Turino, Italy.
- Vroeg. (2019). Gesprekstool effectief hulpmiddel bij verbeteren communicatie van TOS-kinderen. *Vroeg Magazine, 4.* Retrieved from https://www.vakbladvroeg.nl/gesprekstool-effectief-hulpmiddel-bij-verbeteren-communicatie-van-tos-kinderen/

113

Part 2:
Connecting System Levels

"Effective design interventions need to engage at different system levels simultaneously, enabling cross-level interactions between local interventions and institutional frameworks."

Jet van der Touw, Anja Overdiek

7. Leverage Points Revisited: How Social Designers (Can) Apply Systemic Knowledge to Make Impact

DOI: 10.1201/9781003609766

Introduction

Social design is a growing practice field and expected to create impact on socio-technical systems and society at large. It positions designers to commit themselves to long-term visions and coordinate social change accordingly (Tromp & Vial, 2023). This requires designers to have a good understanding of what impact means, and how social design can make it. This understanding is needed both between different designers that work together and between designers and the organizations that could benefit from social design. Often the language used by designers doesn't overlap with the more organizational language of stakeholders. Especially when discussing long-term visions there is a need for a shared "impact language".

One way to align the different languages is to use a systemic perspective on influencing the social as a designer. To achieve long-term societal impact, designers could more systematically adapt a systemic and multilevel perspective to their process, and thus consciously act at different system levels. Until now, system levels have mostly been formulated in relation to design outcomes (Ceschin & Gaziulusoy, 2016) at the levels of product, product-service systems and sociotechnical system. However, the social design field engages with diverse stakeholders where "social innovation outcomes and system-shifting" (Drew et al., 2022) such as a changed neighbourhood culture or new goals for the national health system and not design outcomes as artefacts like products or product-service systems are their aim. The field still lacks the language to talk about societal system levels that social design activities can target, particularly when evaluating impact amongst themselves and with clients. Social designers could therefore profit from concepts developed in broader systems theory.

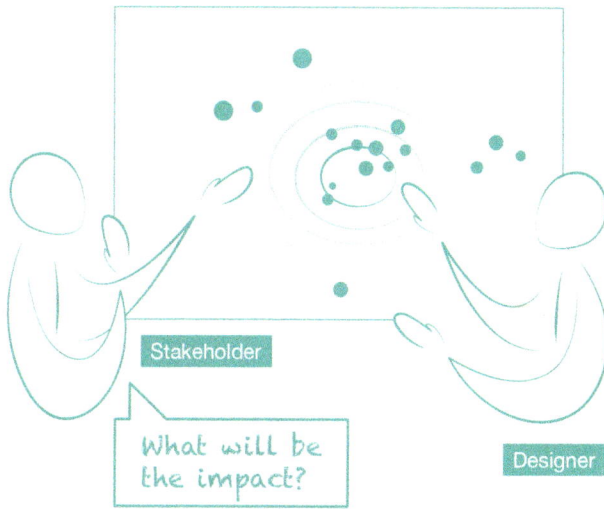

Figure 7.1. Designers and their stakeholders discussing impact

Together with five senior social design practitioners, we formulated this question: "How could systems thinking, particularly leverage points theory, help with this challenge?" In the period of one year, we have researched this topic with the goal of developing a method and language based on Meadow's theory of leverage points in systems (Meadows, D., 1997). In this book chapter we briefly explain the leverage point theory and why we use it, and unpack the different research-through-design steps we engaged in with the five designers. Furthermore, we present the method we developed and demonstrate it in a practice case.

What Is Impact for Social Designers?

At the start of our collaboration, we explored how the five social designers think and talk about the impact of their work today, both the intended impact they wanted to make and the impact they felt they had realized in projects. Different concepts were used to address impact, such as influence, effect, and change. In conversation, they decided that in their experience, impact was the

119

result of many projects together and as such a collaborative goal. For single projects, they rather wanted to relate to their impact as having an "influence with a lasting effect" on a social (sub)system. One of the designers stressed that "impact is not something you can control; it is part of a bigger picture" expressing how they do not work with cause-effect relationships, but with complexity. Another designer mentioned that the notion of "influence" on the other hand, did feel like it was within their reach and something they could control. Consequently, we used "influence with a lasting effect" as the working definition of what we see as impact related to a social design project.

Before entering the phase of co-designing together, there was yet another notion we had to clarify. We noticed that when talking about preparing for or evaluating their impact, the designers predominantly used the word "context" to relate to their application field. An unsafe neighbourhood, an educational program that didn't relate to professional practice, a collaboration to accelerate the transition to fossil-free energy: All fields the designers wanted to make impact on were described as contexts, with people directly involved and stakeholders influenced indirectly. When asked how they planned and monitored their impact on these contexts, they referred to "theories of change" (TOC). In these theories, they confront a present situation with a preferred one, and define the difference as the impact they want to make. Then, they relate this to changes they need to achieve with people directly involved and determine needed outcomes and activities. Of course, assumptions and external influences nurturing and interfering with this TOC are examined. Next to TOCs, the designers also shared their most favoured methods to create impact: designed boundary objects, frame creation method (Dorst, 2015) and generative make sessions (Sanders & Stappers, 2012). None of the designers we worked with had used systems thinking to reflect on their impact, and they were very curious to explore this further.

Theories that Support Our Impact Method

The aim of this project was to see how leverage point theory could offer a base for developing a method for social designers to define and discuss impact. For this we investigated system dynamics and complexity theories. Based on the system dynamics approach, Meadows (1997) came up with a list of 12 "places to intervene in a system" or leverage points, as she calls them. Leverage points are places in complex systems where a small shift may lead to fundamental changes in the "system as a whole" (Meadows, 1999). One such leverage point is, for example, shifting the goal of the healthcare system from treating people who are sick, to preventing people from getting sick, and shifting this goal would change the whole system. Meadow's 12 leverage points were later clustered by Abson et al. (2016) into four broad types of system characteristics out of which the types *design* and *intent* relate most to the social design practice. The leverage points in these two types are e.g. the structure of information flows, power, and self-regulation in a system, but also the goals of a system and paradigms like assumptions about human wellbeing.

 One critique of this way of thinking is that it is too linear and doesn't fit with the interconnected nature of complex systems, particularly during transitions. Considering the transitional state of contexts in which social designers work, we needed a less linear theory to build from. Complexity theory gives a more suitable perspective, focusing on the co-evolution of systems and their environments but offers no practical grip to help define the impact social design wishes to have on a system. Thus, we ended up on a compromise, leaning on the research of Birney (2021) who combines leverage points with living system qualities. She relates places of potential influence (such as where there is *energy for change*) with four of Meadows's leverage points, offering a model for changemakers to find and decide on places to intervene. In the model Birney defines four leverage areas: system structures and resources, patterns of organisations and relations, system goal, collective mentality and paradigm.

121

Co-designing towards an Impact Method Prototype

The method we were developing needed to be simple to use and we didn't want designers and their stakeholders to have to dive deep into systems theory before they could start speaking the same impact language. So, we designed the first prototype of a step-by-step Impact Method. We used Birney's model as a basis for our co-design sessions. In these sessions, we adapted and moulded the model in the context of the designers' work experience with ongoing and completed projects. After each co-design session, we rewrote and redesigned our concept and prototype until it was suitable for use.

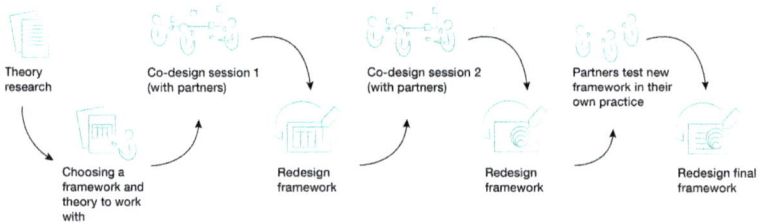

Theory research → Co-design session 1 (with partners) → Co-design session 2 (with partners) → Partners test new framework in their own practice

Choosing a framework and theory to work with → Redesign framework → Redesign framework → Redesign final framework

Figure 7.2. How we did our research together with social designers.

To start discussing societal impact and different system levels at which you can operate and design, you need to know which system we are talking about. While the designers were used to applying the notion of "context" in determining their impact, systems were seen as very abstract concepts. We started with the working definition of a system being a collection of interconnected elements working toward a common purpose. Relevant questions like "where does a system start and end?" and "who and what does a system consist of?" were asked. We noticed that deciding which system to focus on required first to agree with one another what is defined as a system. Only after that could they determine the boundaries within which their project was situated and go into thinking about which part of the system they could tackle. In the Impact Method, the designers first draw circles of different functional systems that connect to their project, such as the healthcare system, the mobility system or the system municipalities work in. This step of the method also brings out the different perspectives of working with a system by defining it on a piece of paper. This leads to a diagram of related systems (to the project or case) and helps those involved to make an informed decision on which system to focus, and exchange about the possibilities to interfere with that system (and, maybe, shift it).

The next step in the method is defining the leverage points within the chosen system. Choosing a particular system like one University or the national health system to tackle, is one thing. Describing how that system works is another. What are its structures (system structures and flows), who is part of it and what are characteristics of the existing relations (patterns of relations and organisations)? What are the goals of this system (system goals) and how do these goals relate to societal paradigms (collective mentality and paradigm)? These are questions inspired by leverage point theory which help define a system and through that envision the possible design interventions with which to make an impact. To sustain designers in addressing these questions, we adapted a "levels of impact" model from Birney (2021) to a visual canvas. Leverage points were reformulated according to how clear they were from the perspective of the practitioners and visual cues of their interconnectedness were iterated (Figure 7.3). For example, the designers expressed a preference to position the four different leverage areas as concentric circles to indicate their nestedness. Some leverage areas are more controllable but have less impact, while others are more difficult to influence but yield a larger effect on the system. Small in this case is relative, as changes in any of these areas will bring about a transformation of the system.

123

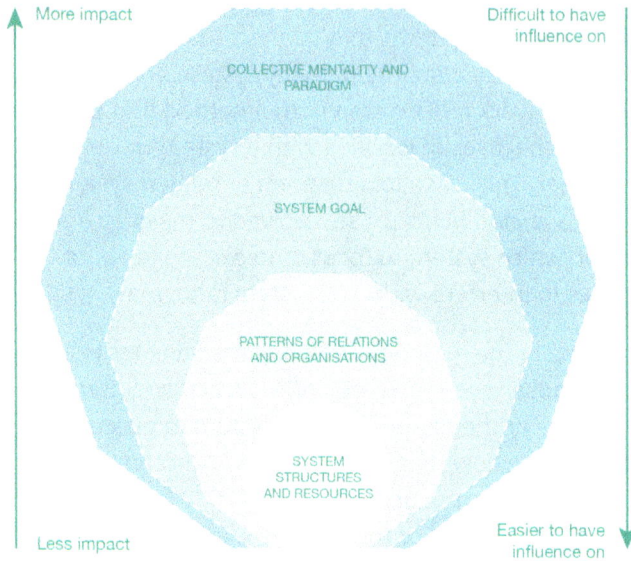

More impact

Difficult to have influence on

COLLECTIVE MENTALITY AND PARADIGM

SYSTEM GOAL

PATTERNS OF RELATIONS AND ORGANISATIONS

SYSTEM STRUCTURES AND RESOURCES

Less impact

Easier to have influence on

Figure 7.3. First prototype of a canvas usable to understand and find leverage areas and their impact in the related system (adapted from Birney, 2021) .

During the last two co-design sessions we worked on the usability of the canvas, employing cases from the designers' practice. We designed a step-by-step guide for this method, with adjusted definitions of the areas and a workflow in three steps (three canvases). One last thing then turned out crucial for the designers. After understanding the context, the system, and its leverage areas, what remained was choosing the focus for the project so that one may start creating and designing interventions. To facilitate this, each leverage area on the canvas was completed with an activating question related to a system-changing strategy. Like that, the three-step method can help social designers and their stakeholders discuss and define the impact they want to make.

Figure 7.4. Social Design Impact Method, consisting of an introduction sheet and 3 templates.

Impact Method in Use: Co-designing "One Government"

As a last step in our research project, we asked the designers to test our finalized Impact Method with one of their ongoing projects. In the below case, the method was used to define the focus for a project's upcoming phase.

In 2024, policy advisors from the Dutch ministry of internal affairs approached the social design firm Zeewaardig with a particular challenge. Nowadays, Dutch entrepreneurs must deal with multiple and very diverse online service systems to exchange and deal with governmental organisations such as the tax office, the institution for handling work opportunities and work-related social security, the chamber of commerce and the national board for entrepreneurs. This creates delays, labour costs for entrepreneurs and an overall dissatisfaction with the government. Taking the lead, and helped by Zeewaardig designers, the policy advisors wanted to embark on a co-design process with entrepreneurs to explore opportunities for a "one government" experience of the future. After a phase of research and interviews, Zeewaardig

125

created a "dancefloor" metaphor for this process, as they noticed that the involved parties did not know each other at all. But after co-design sessions with a group of entrepreneurs and more sessions with the core team consisting of representatives of all the concerned institutions, Zeewaardig reached an impasse. On one side, it was quite easy to come up with design principles for a "one government" experience together with the entrepreneurs; on the other side it became clear that the different government institutions did not know each other's ways of working. Consequently, the back office of such an integrated front would not work. "Our goal was to the 'one government' experience, but that seems very far away. Maybe we should shift our goal. But where?"

Figure 7.5. Zeewaardig working on the 1st step of the Impact Method: Defining systems.

In this situation Zeewaardig called on us. In a session of 1, 5 hours, two designers together with one researcher used the Impact Method. In the first step, they mapped the context with all related organizations on Post-it notes (Figure 7.5). This allowed them to talk about the differences between the partners in the core team as "systems". Some of them were formal systems like governmental organisations, others were more fluid service organisations. The designers realized that their initial goal had been huge: to change several diverse but related social systems! **Their biggest realization in this step was that it would help to see the core team partners as (representatives of) separate systems, related to bigger systems like the Dutch economy or the Dutch welfare state.** They concluded: "Our former goal was so complex that it was no longer transparent. The method helped us to see that our goal is not about one government, but about a simpler government."

In the second step (no figure), the designers walked through the different leverage areas of several of the previously mapped organizational systems. They realized that it was not only at the level of information flows that these systems differ. They also often have different work cultures, other systemic goals such as "rule and regulate taxes" or "support entrepreneurs" and potentially conflicting paradigms (individual rights, entrepreneurship as a value etc.). The designers decided that it would be valuable to visualize and discuss these different systemic constellations together with the core group. This could elucidate possible synergy, but also conflict. In terms of influence, they felt that interventions could be most influential at the relationship level ("getting to know each other's culture"), but also potentially around formulating a shared system goal to add to the existing one. **The method helped them to get a perspective on those areas of the system, they could make the biggest impact on.**

127

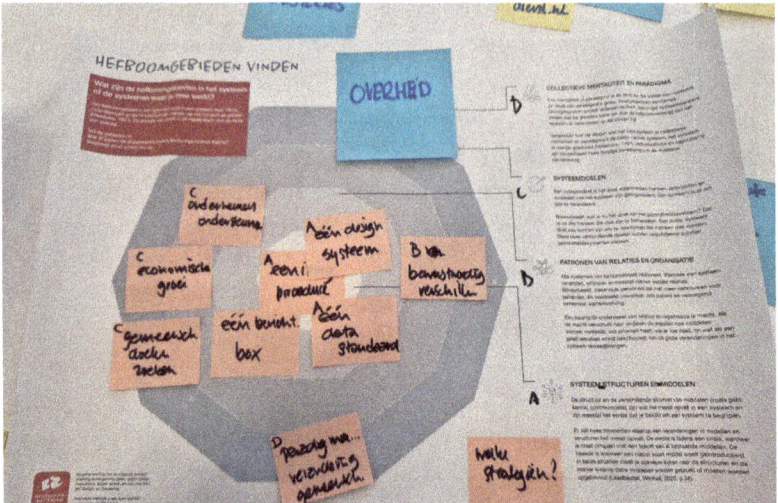

Figure 7.6. Zeewaardig working on the 3rd step of the Impact Method (while using the 2nd canvas) .

In the third step, the designers decided that they wanted to map possible elements on those leverage areas (Figure 7.6) they could influence towards getting to a "simpler government" for entrepreneurs. Next to "one service design, one data standard" (level A: structures & resources), they thought of consciousness-building around differences and shared routines (B: relationships) and a new and shared system goal (C: system goals). They felt that they would not be able to influence paradigms. **Thus, the method helped them to ideate different impactful strategies. "These strategies can be chosen together with the leading parties, intuitively and according to time and resources. They all feel solid because they are grounded in a shareable systemic language", according to Zeewaardig.**

Conclusion

In this co-design project, we wanted to better understand how leverage point theory could be applied to the social design field. We also wanted to develop an alternative perspective on and language for the different systemic levels design needs to tackle to contribute to societal impact and long-term changes. To achieve this, we needed to start from unravelling different perspectives

128

on the notions of "impact" and "system". Translating systems thinking to the context of social design by visualizing leverage points as nested areas in a system proved to be an effective way to bring new perspectives on social design projects and to introduce a shared "impact language" between designers and their stakeholders. The Impact Method we developed together with five social designers can support informed decisions and targeted creativity. It helps with answering which system(s) to focus on, determining the workings of these systems together with the stakeholders, discussing their situated influence on different leverage areas in the system, and ideating on social design activities in these areas. This research took a first step in enriching the work of social designers in the Netherlands with systems thinking and a workable way to introduce it in everyday practice. We hope that, by further iterations in other geographical contexts and publications, it will find its way to professionalize the social design field, yielding much-needed impact on socio-technical systems and society at large towards broader sustainability transitions.

Key Takeaways

The most important insights from this chapter are:

- Defining target systems and system areas to apply design creativity can help to get an advanced perspective on the opportunities of making an impact.
- In addition to working with mapping the context of design challenges, determining influence and mapping interventions on leverage areas can help social designers and their stakeholders to reformulate goals during their projects and thus be more adaptable and achieve more impact.
- The perspective and language of leverage areas in systems helps social designers to evaluate their impact and clearly communicate it to clients.

References

- Abson, D. J., Fischer, J., Leventon, J., Newig, J., Schomerus, T., Vilsmaier, U., von Wehrden, H., Abernethy, P., Ives, C. D., Jager, N. W., & Lang, D. J. (2017). Leverage points for sustainability transformation. *Ambio, 46*, 30–39.

- Birney, A. (2021). How do we know where there is potential to intervene and leverage impact in a changing system? The practitioner's perspective. *Sustainability Science, 16*, 749–765.

- Ceschin, F., & Gaziulusoy, I. (2016). Evolution of design for sustainability: From product design to design for system innovations and transitions. *Design Studies, 47*, 118–163.

- Dorst, K. (2015). *Frame innovation: Create new thinking by design.* MIT Press.

- Drew, C., Robinson, C., & Winhall, J. (2021). System-shifting design: An emerging practice explored. In *Proceedings of Relating Systems Thinking and Design Symposium, RDS11.* Retrieved from https://rsdsymposium.org/system-shifting-design-an-emerging-practice-explored.

- Meadows, D. (1997). *Leverage points: Places to intervene in a system.* Donella Meadows Institute.

- Meadows, D. (1999). *Leverage points: Places to intervene in a system.* The Sustainability Institute.

- Sanders, E. B. N., & Stappers, P. J. (2012). *Convivial toolbox: Generative research for the front end of design.* BIS.

- Tromp, N., & Vial, S. (2023). Five components of social design: A unified framework to support research and practice. *The Design Journal, 26*(2), 210–228. https://doi.org/10.1080/14606925.2022.2088098

Elise van der Laan, Agnes Evangelista,
Kim Poldner, Marjolein Mesman

8. Harnessing Retail Staff as Sustainable Fashion Influencers: A Pathway to Systemic Societal Impact through Applied Design

DOI: 10.1201/9781003609766

Introduction

The fashion industry is facing major sustainability challenges, with environmental degradation continuing due to resource-intensive production processes, making its impact extensive. Consumers increasingly attach importance to environmentally friendly products, and companies are recognizing the importance of sustainability for their long-term viability (Sarokin & Bocken, 2024). Sustainable fashion encompasses practices that reduce environmental impact and promote social responsibility throughout the fashion supply chain (Henninger et al., 2016). Multi Brand Fashion Retailers (MBFRs) stand at a unique vantage point to influence consumer behaviour and engagement with sustainability as they can interact with consumers on a first-line basis (Grewal et al., 2017; Overdiek, 2019). But the main question is how they can use this unique positioning to contribute to the transition to a more sustainable fashion industry?

This chapter shares results from a study that applied design thinking in living lab (LL) settings to explore how retail staff of MBFRs can be supported in fostering sustainable fashion engagement. Retail staff members directly advise customers about their fashion and styling choices and thus they can have a big impact on the customer's behaviour (Hughes et al., 2019; Overdiek, 2019). Even though these interventions play out on an individual level, imagine what could happen when all retail staff in the world can positively influence their customers to make more sustainable fashion purchases. Such micro-level interventions could drive broader systemic change (Cunningham & Jenal, 2016) within the fashion industry.

Therefore, the main research question is tackled from a trilateral systems-level approach. First, the study tries to facilitate a solution on the micro level of the interaction between staff and customer. Research shows that there is a green elephant in the room: MBFR staff want to sell more sustainable fashion brands and customers want to purchase sustainable brands, but they do not dare to talk about it with each other (DSFW, 2017). Second, at the meso level, MBFRs that want to shift to sustainability face barriers in implementing their strategy into daily operations (Retail Innovation

Platform, 2020). Third, in providing a pathway to sustainability engagement that can be emulated for other MBFRs, the study aims to tackle the broader, macro problem, of the fashion industry's environmental impact.

Methods

To answer the main research question, we adopted Living Labs (LLs) as our design intervention. Although no overarching definition of LLs exists, they are commonly referred to as "user-centred open innovation eco-systems based on [a] systematic user co-creation approach, integrating research and innovation processes in real-life communities and settings" (ENOLL, 2022; Steen & Van Bueren, 2017). This form of applied design research was chosen as it emphasises the iterative process of designing, testing, and refining interventions in real-world settings, ensuring that solutions are both effective and adaptable (Overdiek, 2021). LLs have recently been explored as a framework to help facilitate societal transitions (Broek et al., 2020), which is why we adopted this approach to study how we can transition towards a more sustainable fashion industry. When LLs aim to impact a broader level of transition impact, longitudinal perspectives are critical, which includes experiments aiming to create replication and circulation (Broek et al., 2020). In designing a study trying to address broader problems it is therefore important to create learning that can be handed over or emulated (Ballon & Schuurman, 2015; Mastelic et al. 2015; Bergema & De Lille, 2022). Essentially, research generated from a LL should serve as a 'springboard' that other practitioners can build off (Haraway, 2010). This is precisely what we aim to obtain from our LL approach: a pathway that can be emulated for the micro- and meso-level issues outlined above, ultimately helping to support the broader macro level of transition. The following sections elaborate on how we set up and executed our LLs to achieve this aim.

Our study took place in four phases (see Figure 8.1) and was carried out with a consortium of partners. Nine MBFRs partook, all of which were SMEs that served as partners in the study. The Hague University of Applied Sciences (THUAS) led the research project while TMO Fashion business school collected data and surveys. In addition, three fashion/retail branch organizations served first as advisors and then disseminators of the knowledge that was created with this study. The tools we developed as a result of this study were spread through their networks, enabling dozens of retailers to support the transition to a more sustainable fashion industry.

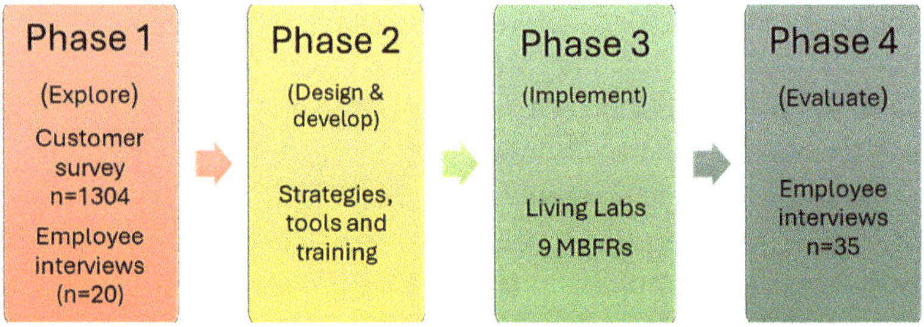

Phase 1	**Phase 2**	**Phase 3**	**Phase 4**
(Explore)	(Design & develop)	(Implement)	(Evaluate)
Customer survey n=1304	Strategies, tools and training	Living Labs 9 MBFRs	Employee interviews n=35
Employee interviews (n=20)			

Figure 8.1. Method phases.

In the first phase, a survey was designed with questions regarding customers' attitudes towards sustainable shopping behaviour. Retailers invited customers to participate via newsletters and physical calls to action at cash registers in stores. The survey was completed by 1, 304 respondents, providing quantitative data on a sample (N = 1304) that is reasonably representative of the customers of SME MBFRs. Furthermore, qualitative data was obtained through interviews, which were held with 20 employees from a total of three participating retailers about their attitudes and behaviour regarding sustainability. Based on this phase, the second phase was used to design tools and strategies that were co-created with students, researchers and MBFRs. The third phase of the study was the implementation of living labs (LL). The final phase occurred after the culmination of the LLs. It was comprised of interviews and surveys conducted with employees of the nine SME partners. The surveys were structured (n=35) and supplemented with several in-depth interviews.

The LL methodology in phase 3 followed seven key steps outlined by Overdiek et al., 2021 (see Table 8.1). The first five steps focused on setting up the LL, and the last three steps were centred around data collection and analysis (Overdiek, 2021). Steps 1 to 5 revolved around securing grants, appointing a coordinator, setting geographic parameters and engaging stakeholders with nine Dutch MBFRs participating to ensure diverse retail environments for data collection. Once the LL had been established, the second set of steps revolved around data collection and analysis. In step 6, qualitative data was collected through non-participatory observations in (physical) stores on predetermined test days (at least 2 days per store) based on an observation form, where the behaviour of customers and employees when using the tools was recorded. We supplemented this with short interviews with customers of participating retailers on the shop floor, a panel discussion conducted once with customers and once with employees, as well as in-depth interviews with employees. We also organized co-creation sessions with the SMEs to garner their input on sustainability-related challenges and barriers as well as possible tools that could be utilized. The SMEs were involved in the design and selection of the tools as well as in methods such as in-store observations, employee and customer surveys. This approach ensured that direct engagement and feedback from key stakeholders was facilitated. Furthermore, it provided a unique data set that could be analysed. Step 6 also ensured validity through an iterative development and testing phase. This was executed by employing iterative cycles focused on tooling effectiveness and staff impact, allowing for the refinement of strategies based on real-time feedback and observations. After transcription of the interviews, we used Atlas.Ti to code the data using the GIOIA method (Gioia et al., 2013). This structured process of data analysis was done in two rounds of coding. First-order coding, where terms and phrases were used to create initial codes, and second-order coding, where we interpreted and refined these codes into overarching themes that derived perceptions and patterns into the effectiveness of SF engagement strategies. This approach ensured a rigorous examination of the data, capturing both emergent insights and their broader implications in SF engagement. Finally, step 7 consisted of the dissemination of the results through our SME partners, social media and the branch organizations for wider reach and impact.

1. Theme	How can MBFRs use their unique positioning to contribute to the transition to a more sustainable fashion industry?
2. Other parties	The Hague University of Applied Sciences, TMO Fashion Business School, Cube Retail and branch partners Modint, Inretail and EK (formerly Euretco), Wagenaar Mode, Modehuis Blok, Van Tilburg Mode & Sport, Berden Mode, Castelijn Mode, Hype Heroes, Rinsma Modeplein, BrandMission, KokoToko
3. Create support	RAAK MKB grant from SIA. 2 years funding duration
4. Lab coordinator	Senior researcher (PhD) and project manager
5. Location	9 locations of MBFRs from different cities in the Netherlands
6. Co-creation in lab	Observations conducted in store. Employees reporting through surveys. Customer surveys.
7. Disseminate information	White papers (and), toolbox and book contribution to be able to reach all target audiences.

Table 8.1. Living Lab Implementation steps, adapted from Overdiek et al., 2021.

Results

The study provided clear insights into the interrelated barriers and drivers of sustainability engagement in MBFRs.

Tools

The iterative process of co-creation led to designing a set of tools that could be utilized to encourage employees interacting with customers to drive SF engagement. First, there were hangers positioned in between the clothes (Figure 8.2), second there were stickers on the floor and on mirrors as well as shelf talkers (Figures 8.3–8.5). Each of these contained QR codes which customers could scan to learn more about the sustainability of certain textile materials. The aim of this toolkit was to lower the threshold for engaging in SF discussions by making it easier for employees to start conversations with customers and vice versa. Although the toolkit had relative success, cultural and behavioural change is needed to get employees on board. Observations show that only 11% of customers read text on the hangers and only a further 2% proceeded to scan the QR code. It was further observed that despite tools being visible on the shopfloor, 90% of employees did not engage in SF conversations.

Figure 8.2. Hangers. Translation of the text on all hangers (top row): 'Do you know what you are wearing?' at top and at bottom 'Turn me around'. Translation for each answer on the hangers from left to right: Thousand liters of water; Wood pulp; A happy sheep; Recycled plastic; Someone's work.' Other side of the hangers (bottom row) translates to 'Scan me! Ask us about it! And scan the QR code for more information. Or go to the website www.nextfashionretail.nl.'

Figures 8.3, 8.4, and 8.5. Shelf talkers and mirror stickers. Translation of the text: Left image 'More sustainable choices - where do you start? Ask us! Follow the tape and scan the QR codes'. Centre image 'Viscose. Do you know what you are wearing? Woodpulp! Scan the QR-code for more information'. Right image 'Do you know what you are wearing? Scan me!'

Organization Support

The best enabler of micro change is through meso-level organizational support like training. Therefore, to increase tool engagement, training and engaging employees in driving SF conversations were critical. Employees and customers were surveyed and observed in the LL setting. The results highlight a clear barrier at the micro level. Many employees think that customers don't want to purchase SF or to engage in conversations around this topic. In reality, customers are more willing to engage in conversations on SF, with surveys from research phase 1 indicating that 80% would want to learn more about sustainability.

In one location where employees indicated an understanding of the need for SF research, they were more likely to engage with the tools and customers. In locations where employees confessed they did not always understand the relevance of the study, they also did not immediately see how the tools could make it easier to talk about SF with customers. It was further observed that intrinsically motivated staff regarding sustainability in general had far better success with customers. Many employees indicated a desire to use the tools, but in practice not all understood how to use them, and many did not follow up in looking into their use—with only 28% visiting the website that provided more information about the tools. However, this could be attributed to the fact that employees who received training/explanation about the project were not always the same employees that worked during the days that we tested the tools on the shop floor. The survey results from phase 4 show that approximately 25% of employees lacked explanation about how to use the tools. We found it is crucial to have training by top management to ensure that employees were informed as this was more effective than providing staff with do-it-yourself training kits or external trainers explaining how the toolkit works.

Supporting Drivers

Another driver includes antecedents such as competitions or promotions (Figure 8.6). Many employees reported feeling that giveaways were fun and useful at the checkout as they would lead directly to customer engagement in SF discussions. Furthermore, the Amsterdam location used employee-organized challenges during the test period to motivate each other in engaging customers with the tools as much as possible. Consequently, promotions provide a valuable mechanism for driving employees to engage with customers in SF discourse.

Figure 8.6. A giveaway promotion. Moth repellent for clothes. Translation of the text (left side): 'You can hang this ring and card on your cloth hanger as a moth repellent. The scent from the cedar wood will keep moths out of your wardrobe. Do you want more tips on how you can best take care of your clothes? Scan the QR code! Or go to the website'. Right side 'You receive this ring and card as a gift from Rinsma'. Rinsma was one of the retail partners in the project.

Albeit there was some interest in the tools, little interaction between customer and employee regarding the topic prevailed. Phase 3-4 therefore shows that tools, although helpful, have a maximum impact only if there is additional employee training provided.

Discussion

Tools

The initial deployment of tags and QR codes as tools to promote SF in MBFRs demonstrated minimal effectiveness without the active engagement of employees. Employee hesitation to engage with sustainable fashion (SF) tools may stem from a lack of confidence, insufficient knowledge about sustainability, or uncertainty about how customers might respond to such discussions. Many employees fear that broaching SF topics might disrupt the sales process or face customer disinterest. Additionally, customer engagement with SF tools was relatively low, likely because the tools relied heavily on passive interaction, such as reading or scanning QR codes, which require active effort on the part of the user. This finding is supported by Takens et al. (2022), who note that only a small fraction of consumers use QR codes, a challenge compounded by the demographic profile of customers in these retail settings (Albăstroiu & Felea, 2015). The subsequent phase highlighted the necessity of involving employees more integrally, underscoring their role in transforming the SF landscape within MBFRs. Literature suggests that the true potential of green tags and QR codes (Grewal et al., 2017) is unlocked only when combined with meaningful customer interactions facilitated by staff, as these interactive experiences significantly influence consumer behaviour (Hollebeek, 2011; Sonnenberg et al., 2014; Henninger, 2015; Hwang et al., 2015). Our results contribute by indicating that employees must see the importance of the cause and perceive that customers are open to these discussions. Literature indicates that when employees, in a retail setting, receive basic training that leads to a superficial change in how they present themselves, but not "modifying their true feelings" or attitude, customers will sense the lack of authenticity and it will have a negative effect on consumer behaviour (Tremblay, 2023). This further consolidates the critical importance that training should comprise of education as to why SF engagement is important and necessary. Management training was critical in this regard: management can receive a train-the-trainer session and offer easy-to-use microlearning courses to their staff that mainly focuses on behavioural development which leads to increased knowledge of tool use and its importance. Providing management coaching further helped translate the mission into fostering an organizational culture centred around sustainability. This clearly highlights the necessary interplay in ensuring that the meso level contributes toward motivating the micro level.

Another meso-level strategy that could have an impact on the micro level of how staff and customers interacted with the tools and sustainability, was in offering rewards for participation. Through using one store (Amsterdam), to initiate a challenge among employees based on tool engagement, the potential of promotions to generate excitement and active participation in sustainability was revealed. Employee attitudes towards SF significantly influence tool performance and consumer engagement within MBFRs. Studies show that brands benefit when employees have strong relationships with them, particularly in the context of sustainability (Hughes et al., 2019; Khandual & Pradhan, 2019; Sarokin & Bocken, 2024). Our results support this and further highlight that intrinsically motivated staff, who understand and are committed to SF principles, are more effective in connecting with customers. Ultimately, this leads to enhancing consumer engagement with sustainability at the micro level.

Map to Broader Impact

The connection between micro- and macro-level impacts is critical, as micro-level interventions often serve as the building blocks for systemic change. For example, in the energy sector, household adoption of solar panels, initially supported through localized incentives, has led to significant shifts in national energy policies and infrastructure investments (Rynska, 2022). This underscores how well-designed, localized interventions—when scaled or replicated—can catalyse macro-level transformation. By addressing micro-level interactions, such as equipping retail employees with tools to engage customers on sustainability, the fashion industry can lay the groundwork for broader systemic change. Managers indicated that they found it difficult to disseminate tool use and engagement in larger stores, but smaller stores where more individualized hands-on training and motivational coaching could take place had more success. Thus, there is a need to place considerable effort into staff training and engagement with tools for a small and local design intervention to have the most profound level of effect in supporting societal transition to a more sustainable fashion industry. This highlights employees as the key drivers of sustainable change, in line with previous literature from other business settings, which affirms that employee perception and relationship to sustainability initiatives directly corresponds with their involvement (Lee et al., 2013). Success of tools at the micro level hinges on training interventions at the meso level. This will serve to get customers to engage with conversations on SF and begin identifying with sustainability norms and values. Finally, to enhance SF engagement within MBFRs, a long-term orientation is crucial, which is also reflected in literature for living labs aimed at supporting transitions (Broek et al., 2020). Embedding SF into the fabric of MBFR operations encourages the development of subjective norms around sustainability among staff and customers alike. This approach not only educates customers on the value of sustainability but also fosters a retail environment where SF brands thrive, thereby providing a strategy that can be emulated across MBFRs.

Future research and practitioners should, however, consider that in order to create this long-term impact and transition, there are a series of paradoxes that must be navigated such as informing consumers about negative impacts within a purchasing environment (Han et al., 2024; Strübel et al., 2023). The challenge is to do so in a way that doesn't deter purchasing or conflict with the immediate commercial interests of the store. Retailers must navigate the tension between working with the non-sustainable brands and promoting a broader shift towards sustainability through organisational and cultural change (Feeney et al., 2023; Yang et al., 2017). Future applied design research within MBFR settings should use the findings outlined above as a springboard for future research that further manages these paradoxes. This will further ensure these solutions are effective and adaptable across different levels of the retail environment to maximise the macro impact of transitioning the fashion industry towards sustainability. However, it is important to note that this study is limited due to the limited timeframe in which the LL could be conducted. We recognize that the LL would have benefited from a longer-term timeframe than the scope and funding of the project enabled us. As such, another avenue for future research could be to embark on a longer study of a similar nature that also addresses the above tensions.

Conclusion

In exploring how MBFRs can use their unique positioning to contribute to the transition to a more sustainable fashion industry, we created a series of interventions aiming to reduce the barriers for MBFRs in creating sustainability engagement with their consumers. Even though the interventions created were carried out on an individual level, if retailers across the Netherlands adopt them to help support more sustainability engagement, this micro-level intervention could help drive a broader level of systemic change in the fashion industry. Our small-level design intervention was the creation of tools that we tested at several living labs. These tools are now available for all retailers in the Netherlands, supporting them to adopt more sustainable practices such as selling sustainable fashion brands. We could as such argue that we have performed a small-scale intervention on a micro level that is able to influence organisational strategy and cultures (meso level) and can possibly have a ripple effect in the transition towards sustainable fashion (macro level). Further research is however needed to understand how a relatively small-scale, temporary, and local design intervention such as ours can continue to have a broader influence on long-term societal change processes and transitions.

Key Takeaways

The findings of this study offer actionable insights for designers and design researchers, highlighting the role of localized, user-centred interventions in addressing complex challenges and their potential to drive broader organizational and societal change.

- Design user-centred interventions: Involve users and stakeholders in an iterative design process to ensure interventions are tailored to their needs, behaviours, and cultural contexts. This helps create solutions that are functional and also resonate on an individual level.
- Embed contextual and organizational insights into design: Tailor tools and interventions to fit the specific workflows and constraints of the context (e.g., small vs. large organizations), ensuring smooth integration into existing systems.
- Leverage diverse dissemination methods: Use design principles that prioritize simplicity and accessibility for broader adoption. Create designer-friendly toolkits, case studies, and visual documentation to share findings with the broader design community, enabling collective learning and fostering larger systemic change.

146

References

- Albăstroiu, I., & Felea, M. (2015). Enhancing the shopping experience through QR codes: The perspective of the Romanian users. *Amfiteatru Economic Journal, 17*(39), 553–566.

- Bergema, K., & De Lille, C. S. H. (2022). Scaling up: From labs to systemic change. In *Proceedings of Relating Systems Thinking and Design, RSD11*. Retrieved from https://rsdsymposium.org/scaling-up-from-labs-to-systemic-change/

- Ballon, P., & Schuurman, D. (2015). Living labs: Concepts, tools, and cases. *Info, 17*(1), 1–11.

- Broek, J. van den, Elzakker, I. van, Maas, T., & Deuten, J. (2020). *Voorbij lokaal enthousiasme: Lessen voor de opschaling van living labs.* Den Haag.

- Cunningham, S., & Jenal, M. (2016). *Rethinking systemic change: Economic evolution and institutions.* Technical Paper. The BEAM Exchange.

- DSFW. (2017). De consument verwacht beter geïnformeerd te worden door de retailers en modemerken. *Dutch Sustainable Fashion Week.* Retrieved September, 2020, from http://www.dsfw.nl/persbericht-weinig-consumenten-letten-op-verantwoorde-productie-bij-kopenvan-kleding

- EnOLL. (2022). What are living labs - European Network of Living Labs. *European Network of Living Labs.* Retrieved from https://enoll.org/about-us/what-are-living-labs/

- Feeney, M., Grohnert, T., Gijselaers, W., & Martens, P. (2023). Organizations, learning, and sustainability: A cross-disciplinary review and research agenda. *Journal of Business Ethics, 184*(1), 217–235.

- Gioia, D. A., Corley, K. G., & Hamilton, A. L. (2013). Seeking qualitative rigor in inductive research: Notes on the Gioia methodology. *Organizational Research Methods, 16*(1), 15–31. https://doi.org/10.1177/1094428112452151

- Grewal, D., Roggeveen, A. L., & Nordfält, J. (2017). The future of retailing. *Journal of Retailing, 93*(1), 1–6.

- Han, J., Woodside, A. G., & Ko, E. (2024). *Does consumer knowledge about sustainable fashion impact intention to buy? Asia Pacific Journal of Marketing and Logistics*, 36(10), 2390–2410. https://doi.org/10.1108/APJML-05-2023-0456

- Haraway, D. (2010). When species meet: Staying with the trouble. *Environment and Planning D: Society & Space, 28*(1), 53–55. https://doi.org/10.1068/d2706wsh

- Henninger, C. E. (2015). Traceability: The new eco-label in the slow-fashion industry?—Consumer perceptions and micro-organizations' responses. *Sustainability, 7*(5), 6011–6032.

- Henninger, C. E., Alevizou, P. J., & Oates, C. J. (2016). What is sustainable fashion? *Journal of Fashion Marketing and Management: An International Journal, 20*(4), 400–416.

- Hollebeek, L. (2011). Exploring customer brand engagement: Definition and themes. *Journal of Strategic Marketing, 19*(7), 555–573.

- Hughes, D. E., Richards, K. A., Calantone, R., Baldus, B., & Spreng, R. A. (2019). Driving in-role and extra-role brand performance among retail frontline salespeople: Antecedents and the moderating role of customer orientation. *Journal of Retailing, 95*(2), 130–143.

- Hwang, C. G., Lee, Y.-A., & Diddi, S. (2015). Generation Y's moral obligation and purchase intentions for organic, fair-trade, and recycled apparel products. *International Journal of Fashion Design, Technology and Education, 8*(2), 97–107.

- Khandual, A., & Pradhan, S. (2019). Fashion brands and consumers' approach towards sustainable fashion. In *S. S. Muthu (Ed.), Fast fashion, fashion brands and sustainable consumption* (pp. 37–54). Springer Singapore. https://doi.org/10.1007/978-981-13-1268-7_3

- Lee, E. M., Park, S. Y., & Lee, H. J. (2013). Employee perception of CSR activities: Its antecedents and consequences. *Journal of Business Research, 66*(10), 1716–1724.

- Mastelic, J., Sahakian, M., & Bonazzi, R. (2015). How to keep a living lab alive? *Info, 17*(4), 12–25. https://doi.org/10.1108/info-01-2015-0012

• Overdiek, A. (2019). *Lokale helden in de retail: Wie zijn ze en wat hebben ze nodig?* Utrecht: Publicatie Retailagenda.

• Overdiek, A., Geerts, H., & De Lille, C. (2021). Innovating with labs: That's how you do it! Insights from future-proof retail. Hague University of Applied Sciences. ISBN: 978-9083078021

• Rynska, E. (2022). Review of PV solar energy development 2011–2021 in central European countries. *Energies, 15*(21), 8307.

• Retail Innovation Platform. (2020). *Met grote sprongen & kleine stappen naar een circulaire retail: Welke bijdrage aan de kennisontwikkeling binnen dit thema willen we leveren aan de retailsector?* Position Paper, March 2020. Retrieved from https://www.retailinsiders.nl/docs/ab891019-1c12-4868-8a8a9e32c3859c50.pdf

• Sarokin, S. N., & Bocken, N. M. P. (2024). Pursuing profitability in slow fashion: Exploring brands' profit contributors. *Journal of Cleaner Production, 444*, 141237.

• Strübel, J., Goswami, S., Kang, J. H., & Leger, R. (2023). Improving society and the planet: Sustainability and fashion post-pandemic. *Sustainability, 15*(17), 12846.

• Sonnenberg, M., van Zijderveld, V., & Brinks, M. (2014). The role of talent-perception incongruence in effective talent management. *Journal of World Business, 49*(2), 272–280.

• Steen, K., & van Bueren, E. (2017). The defining characteristics of urban living labs. *Technology Innovation Management Review, 7*(7), 21–33. Retrieved from http://timreview.ca/article/1088

• Takens, J. et al. (2022). Wat vinden Vlaamse consumenten van (de communicatie over) duurzame kleding? *[PowerPoint slides].* Creative and Innovative Business, Thomas More Hogeschool.

• Tremblay, M. (2023). Is authenticity needed in service-sales ambidexterity? Examination of employees and customers' responses. *Journal of Personal Selling & Sales Management*, 1–21.

• Yang, S., Song, Y., & Tong, S. (2017). Sustainable retailing in the fashion industry: A systematic literature review. *Sustainability, 9*(7), 1266.

149

Poorvi Garag

9. Unravelling Lifeworlds: Farmers and Consumers of Black Pepper

DOI: 10.1201/9781003609766

Introduction

As a contribution to the understanding of the postcolonial capitalist international commodity trade system and applied design research theory and practice, this chapter works at unravelling the socio -economic lifeworlds of the farmers and consumers of black pepper. The sites of ethnographic research are supermarkets in Eindhoven where Dutch consumers buy their pepper and the farms in Shimoga where Black Pepper is grown. The Indian farmers and Dutch consumers being the two endpoints of the complex international trade have been chosen to identify mechanisms at work in the trade across countries through a subjective lens. This design research jumps between my personal experience, primary research and secondary research lying at the intersection of personal, interpersonal, and national subjective experiences.

The ethnographic study's findings, combined with critical fabulation methods creates a social design project that aims to provoke reflection and curiosity. As a contribution to a critique of the postcolonial capitalist international commodity trade system, this chapter works through the lenses of ethnographic design research, eventually producing a social design project titled "Peperduur: As Expensive as Black Pepper.

Keywords

Critical Ethnography; Post-colonial; Social Design; International Trade Development; Critical Fabulation; Applied Design Research

Definitions

Social Design: Ideo defines Social Design as a process that a) encourages community facilitation, b) is supportive and empowering for those involved, c) offers an innovative and feasible process, d) does not try to change people's behaviour, e) draws on cultural traditions and beliefs to frame problems within society, and lastly, f) acknowledges the importance of the wider influence of design, for example, upon the environment (IDEO, 2015).

Ethnographic research: Ethnography can be defined as the 'first-hand experience and exploration of a particular social or cultural setting on the basis of (though not exclusively by) participant observation. Observation and participation (according to the circumstance and the analytic purpose at hand) remain the characteristic feature of the ethnographic approach' (Atkinson et al., 2001).

Neocolonialism: Neocolonialism is a new form of subjugation affecting the economic, social, cultural, and political life of formerly colonised countries. It manifests as economic imperialism, globalisation, cultural imperialism, and conditional aid, aiming to influence or control developing countries without direct military or political dominance (Nkrumah, 1965).

Post-colonialism: Postcolonialism is a studied engagement investigating the experience of colonialism and its past and present effects, both at the local level of ex-colonial societies and at the level of more general global developments (Said, 1978).

Critical fabulation: Critical fabulation is a literary method by Sadiya Hartman is a mode of historical writing that is both creative and semi-nonfiction as a means of redressing history's omissions, particularly those in the lives of enslaved people (Hamer, 2020).

Speculative fabulation: Speculative fabulation is a wide set of methods and techniques rooted in everyday storytelling practices through which people can interact with imaginary worlds to create imaginary realities radically different from the one we know, through storytelling and reflective thinking (Harraway, 2016).

Social Design

This chapter reveals the continual construction of my identity as an upper middle class Kannadiga woman social designer of 2024, in a postcolonial-capitalist world. As an international social designer, I've blended my experiences in Indian and Dutch social design, which has deepened my understanding of social design as an optimistic, research-driven approach to addressing complex societal challenges. In social design, insights and learnings from

153

extensive research are applied through various design forms—ranging from products and services to frameworks, methods, campaigns, and installations—all to drive positive societal change. Social design has evolved over the decades, with roots in design research theory and practice, formulated of writings from Victor Papanek, Nigel Whiteley, Victor Margolin, Arturo Escobar, László Moholy-Nagy and more contemporaneously, Tim Brown's Design for Social Impact and many more countless contributions (Chen, 2016). Ultimately, social design challenges us to reflect and reimagine the society around us, it expands on design research practices, ideation and production to design beyond objects, establishing that design influences, in many ways, the design of society, which in turn influences the society of design.

Colonialism, Neo-colonialism and Postcolonialism

The theoretical foundation of this project is built upon understandings of existing discourse and studies on postcolonial-capitalist investigations of the trade between the global south and north. My research in postcolonial thinking is understood from Homi Bhabha's theory of hybridity and Gayatri Spivak's deconstructive theorising of the subaltern (Bhabha, 1994a, 1994b; Spivak, 2010). I understand postcolonialism as a studied engagement investigating the experience of colonialism and its past and present effects, both at the local level of ex-colonial societies and at the level of more general global developments, the study of which was pioneered by Edward Said and his book "Orientalism" (Ivison, 1978). One foundational step before postcolonial theory was the widespread neocolonial thinking defined by Sartre as the establishment where former colonisers maintain colonial ties with their former colonies through economic dependencies, the theory of which, Kwame Nkrumah's book Neo-colonialism: The Last Stage of Imperialism discusses at great length (Nkrumah, 1965). Finally, integral to the research about black pepper international trade in a postcolonial capitalist world is Karl Marx's initial theories on alienation and the impact of the capitalist society on humans (Marxists Internet Archive, n.d.).

Ethnographic Research

Ethnographic Methodology

Secondary research along with multi-sited ethnography in and around Shimoga, India and Eindhoven, Netherlands are the primary methods for the research produced in this chapter, which is supplemented with reading, writing and observations. Impromptu and/or planned one-on-one conversations and group conversations during largely casual and comfortable situations such as a gathering of friends, shared leisurely activities, conversations in the supermarket, living within the farming community and sharing a meal with farmers are the focal points of the research methodology. The participants of the ethnographic interviews are of these demographics:

1. Eight Indian pepper farmers in Karnataka in and around Shimoga Pepper Farms; additionally, three labourers and other stakeholders/natives of the area

2. Pepper consumers in the Netherlands (fifteen people from my social network and thirty people who answered a survey that I designed and circulated—Black Pepper Survey)

The Indian farmers in Shimoga, India, and Dutch consumers in Eindhoven, Netherlands, being the two endpoints of the complex international trade have been chosen to identify mechanisms at work in the global black pepper trade across countries through a subjective lens.

This research jumps between my personal experience, primary research and secondary research lying at the intersection of personal, interpersonal and national subjective experiences. Taking into consideration pre-British oral history traditions in India, my interpretation and understanding of the research is considered crucial to the delivery and extension of the knowledge to other people and my role as the storyteller is passed from one reader to another (Kambar, 1994). Storytelling here becomes a mutual process, built around not only the narratives but also through the interpretive interaction between me as a researcher and the people I speak with.

155

Unravelling Lifeworlds

On 20th May 1498, the discovery of sea routes from Europe to South Asia by Vasco Da Gama gave Europeans access to India, disrupting Arab-mediated spice trade between India and the rest of the world, leading to the colonisation of the Indian subcontinent for two hundred years (Cartwright, 2021). The Dutch created the VOC in 1602, dominating trade between Asia and Europe (Dyck, 2019). After its fall, India became part of the British Empire, which dismantled local industries through violence and taxes to enforce reliance on British goods (Tharoor, 2016, pp. 41–62). India was relegated to producing and exporting specific crops or minerals, with colonial policies enforcing commodification of labour, unequal wages, and poor working conditions (Prashad, 200, pp. 207–223). Colonial rule in India enabled the oppressive system of food production entailing commodification of labour, unequal wage distribution, poor working conditions and limited rights (Jackson, 2021). Even after independence, neocolonial practices kept India economically dependent, turning its market into a raw material source for former colonisers (Aijtinuk, 2007).

The first among the commodities traded between India and Europe, was black pepper. Grown in intercrop plantations with coffee and areca nut plants and native to India, black pepper is found in high-elevation, dense forests of shade-giving trees like the Indian coral tree and silver oak tree (Subramanian, 2021). In regions like Karnataka and Kerala in the south of India, intense and long rainfall along with moist soil rich with organic matter is conducive to the cultivation of pepper, all of which fill the air with crisp earthy, woody and spicy scents. Eight thousand kilometres away from the plant it has been grown on, black pepper reaches Dutch supermarkets as a commodified product, stripped of its origins and context (Kearney, 2010). Black pepper today in the Netherlands is shipped in transparent cylindrical bottles, stacked in cardboard cases, and unloaded into supermarkets in the Netherlands. Each bottle has a sticker with the product name "Noir Poivre/ Zwarte Peper/ Black Pepper" and the company logo but lacks information about the country of origin or context. Brands like La Drogheria and Apollo use a ship logo, hinting at colonial

156

connotations. Kania's bottle features an artificially generated image of ripe blackberries, which does not represent the actual colour, size and form of the black pepper plant, further obscuring the pepper's origin and reducing the product to barcodes and SKUs. The two major brands (Verstegen and La Drogheria) are selling black pepper in the Dutch supermarkets Albert Heijn and Jumbo, and both brands are similarly priced, with Verstegen costing 2.99 Euro for 50 grams and La Drogheria's black pepper costing 3.29 Euro for 47 grams. La Drogheria also offers another black pepper variety labelled as "Tellicherry Black Pepper", which is also priced at 3.29 Euro for 47 grams and is the only black pepper sold in the major Dutch supermarkets that mentions the origins of the raw materials to be India (Verstegen, n.d.; Albert Heijn, n.d.).

In a conversation *with multiple black pepper farmers,* while I was sharing a cup of tea with them, this conversation took place (George, personal communication, January 11, 2024; P. Narona, personal communication, January 11, 2024).

"P: "See this, this is black pepper from Netherlands. The price of this is three hundred rupees for 50 grams."*

G: "What?! Three hundred rupees? Just for this much, three hundred rupees? If you gave us six hundred rupees we could give you a kilogram of pepper!"

P: "You have to turn it like this." (I said, showing them that the top of the grinder must be turned in order to use the pepper inside.)

G: "And what is this? They have labelled it green?"

P: "That is actually organic, it is called biologisch. They label organic products like that."

G: "Is it so expensive because it is organic?"

P: "Actually the price is not that different, just ten cents more, that is ten rupees more. Do you practise organic farming?"

157

P: *"Actually, people promoting organic farming were here, they explained everything. But, despite doing the whole process and taking that much effort, do you know at what rate the pepper was sold? Just six hundred and ten rupees per kilogram. Just for ten rupees we have to put in this much effort. The organic pepper requires fertiliser twice more than normal pepper, it requires a lot of effort. We also need labourers who put the fertiliser twice right?"*

Across supermarkets in the Netherlands, the '*biologisch*' (sustainable) variety is costlier by approximately 10%, and while the EU organic laws and policies are strict on the environmental impact, the EU organic strategy including the Farm to Fork strategy mentions little about the social conditions within which produce is grown and harvested in (Appunn, 2021). Marianne Van Kemp, the Chief Sustainability Officer at Verstegen, highlighted that in the EU, food sustainability legislations are taken very seriously. She said, *"You know that in Europe we have the CSRD, the CSDD we have many legislations in Europe coming. Which is exciting and overwhelming and well, you know the discussions about legislation don't have to tell you that. Our products come from all over the world: Indonesia, China, India, of course Turkey, North Africa, Europe, Canada, South America, so all over the world. We have about 300 different herbs and spices which is a lot and you can imagine that it's almost impossible to make all those supply chains completely sustainable. Because, you know, you need to make them transparent, you need to talk not only farmers, but also the in-between, the processors, the collectors. So it's really complicated. So we have to make choices"* (M. Van Kemp, personal communication, January 31, 2024).

Consumers, on the other hand, had something else to say. *"I usually buy the cheaper black pepper, also at Lidl. Like this one"* (she shows me the black pepper powder bottle she picked up at Lidl). *"It is also easy that I don't have to grind it"*, says Lisa, a Dutch consumer (L. Post, personal communication, September 23, 2023). Ten out of thirty respondents to the Black Pepper survey, were not concerned about the source of their cooking ingredients (Garag, 2023). The EIT Food Trust Report that surveyed 20, 326 consumers across eighteen European countries showed that in 2021, less than half of the consumers trust food manufacturers and authorities and only a third believe the food they eat is sustainable. The study also shows that although 76% of Europeans are motivated to live a sustainable life, only 51% of these consumers take sustainability into account while making food choices (EIT Food, 2021). Efficiency and convenience drives the production, distribution and consumption of food often reaching consumers via lengthy, international supply chains beyond individual comprehension (Kneafsey, 2008). Seventeen out of thirty respondents of the Black Pepper survey in the Netherlands were not aware of where black pepper came from, and twenty-five of the thirty consumers of black pepper were unable to pinpoint how the plant would look like illustrating that the separation of the spheres of production and consumption of black pepper via abstract value chains (Cheyney, 2016; Garag, 2023). This evidence of the lack of contextual understanding of commodities within consumers of black pepper reveals that in the world's economy, eventually most commodities are translated and ripped off from their lifeworlds (Tsing, 2015, pp. 64–65).

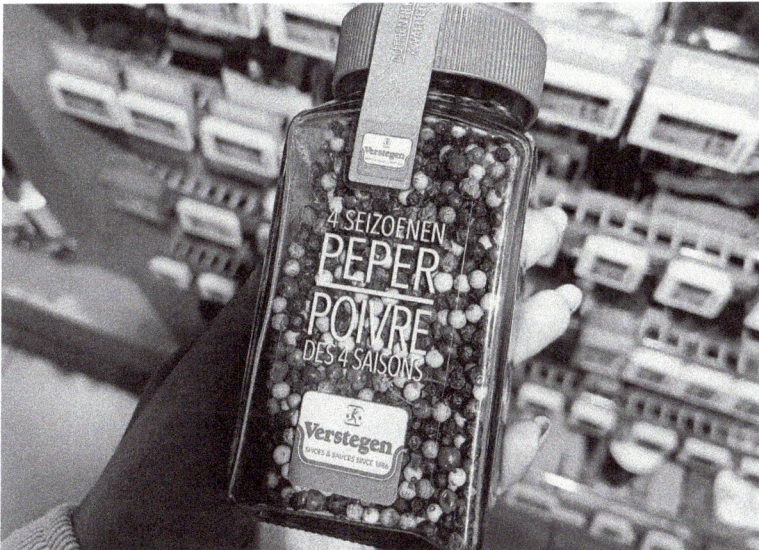

Figure 9.1. Garag, Poorvi. Black pepper jar in Jumbo, Netherlands. September 2023.

My primary contact for pepper farms in Shimoga, Mr. Umesh Rao explained to me during our discussion about the condition of farm labourers and the residual effects of colonialism, that the socio-economic structures of farms in India having changed from the British colonial rule, continue to create socio-economic divides. Pepper prices have been on a steady decrease since 2015/2016 due to steady production increase, putting pressure on the industry, especially smallholders who are burdened with the high costs of production and the low selling prices (Brazspice, 2023; Mazlan, 2020). *"This is how everything is and has been. This is the reality that we live every day. Everything that we eat is second class food because the best of the farms' produce are all exported, why? Because people abroad are willing to pay more for the food. It is part of our daily lives. Since we live this reality, we understand it."* said Col. Ujwal Kammar of the Indian Army (U. Kammar, personal communication, September 16, 2023). Brought up in Karnataka, close to farmlands but having travelled all over India Col. Ujwal is an Indian man in his late forties who introduced me to Mr. Umesh Rao, a farmer in the Shimoga region. In forty acres of Mr. Mark's (a black pepper farmer, known to Mr. Umesh Rao) plantation, almost thirty minutes' drive away from the nearest settlement, isolation and alienation manifests almost physically. With the sun setting, Mr. Mark would release his strong and fierce farm dogs into the fields to protect his estate, himself and his wife. *"Notice how he said he just returned from Koppa? That means he would have gone to Koppa just around a week back itself. In his life, other than labourers and wife, he doesn't have anything else, and of course those two dogs. Life is so difficult. Did you see how he has tamed those dogs? Here, far away only he and his wife stay in this house. His children too are not here, so they have to fend for themselves. The life of plantations is very isolating. What he must be going through mentally we just cannot know. Before the British there wasn't this plantation life,"* said Mr. Umesh Rao (U. Rao, personal communication, January 12, 2024). At the next farm visit, Mr. Prashant Narona, another black pepper farmer, further explained labour issues and the laborious nature of black pepper farming. *"There is a huge labour problem now. Labourers just don't come for work, regardless of how much you pay them. That's not all; if we get six hundred rupees for one kilogram of pepper, we are satisfied, but for us to get that six hundred rupees we need to struggle so much,"* said

Mr. Prashant Narona with great dissatisfaction (P. Narona, personal communication, January 11, 2024). Even for a higher payment than minimum wage market prices, farmers struggle to find labourers to work their farms (Pabolu, 2023). When asked of her struggles in the field, Lata, a labourer in Mr. Prashant Narona's farm responded, *"We don't have any struggles, but this work itself is a laborious struggle. We too have our own farm, we grow rice, coffee and pepper. We don't have a big enough land, so we must work here too"* (Lata, personal communication, January 11, 2024). Earning around 600 to 800 rupees or 6, 69 to 8, 92 euros for a whole day of work, Lata mentions that despite having both her own farmland and working in Mr. Prashant's farm, it is still not enough for them to have what they need.

By researching subjective experiences of the farmers and labourers, I came to realise that socio-economic problems in black pepper farming are deeper than visible unequal wages, social discrimination or lack of recognition for hard work, making the socio-economic landscape of black pepper farming complex and convoluted. The ethnographic research conducted in the black pepper farms and supermarkets in the Netherlands reveals that at both ends of the supply chain, both farmers and consumers experience alienation, lack of information and loss of agency in their own respective ways.

161

Figure 9.2. Garag, Poorvi. Farmers in the black pepper farm, South India. January 2024.

"Peperduur: As Expensive as Black Pepper"

The narratives that were collated in this chapter can only begin to be grasped once the organic nature of this information is understood: testimonies, theory and interpretation of experiences. Through this research it is understood clearly that the implications and consequences of trade extend far beyond the visible statistics of sales and profits and are subjective in nature. The task of recognising the disconnection between production and consumption, and acknowledging the socio-economic and historical complexities embedded in everyday commodities is upon a variety of stakeholders, such as consumers, farmers, policy makers, spice companies, etc. across the world, but especially in contexts where ingredients come from far and colonised lands and people. As a social designer, my approach towards the application of the conducted design research was to ideate on how I could disseminate the understood information and invoke curiosity and reflection among those who would read/see my project.

The extensive reading of history in combination with on ground primary research sparked the need for an appropriate ideation methodology that could combine fact and fable to design a world of alternatives. Sadiya Hartman's critical fabulation approach formulated the creative framework towards the design project. Her writing provokes the line between history and imagination, and upon reading "Venus in Two Acts," I reflected on how much of recorded history is an instance of those in power neglecting the history of those that were oppressed (Hartman, 2008). My project "Peperduur: as expensive as black pepper" takes one historical event and fabulates an alternative history, which then becomes an inspiration to the design of an object and an exhibition setup.

"Peperduur" fabulates an alternative history tangential to documented history in relation to international spice trade. The project works with the central question, "What if black pepper continued to be used as currency?" as a critical speculation of the recorded history of black pepper when it was used as currency for rent, dowry, taxes, etc. to question the socio-economic state of labour-intensive agro-commodities today. The fabulation that I developed centred itself around this short narrative:

"A world where black pepper remains a currency and the value of goods and services fluctuates with pepper prices, influencing global markets and financial systems. For people to use black pepper as a currency, specialised objects, services and systems are designed."

Inspired by the design heritage of Karnataka in South India— the origin of black pepper and the region of my ethnographic research—I designed a crafted object, a 'black pepper wallet,' in response to the fabulation. Along with the 'black pepper wallet' as a design to symbolise the fabulation, a video and a materialised timeline gives context and experience to the object design. The timeline functions as a contextual element to the project, and it includes various key events in history in relation to the international trade of black pepper. The exhibition's timeline, 'black pepper wallet,' and video are designed to provoke a shift in thinking and perception. The design of the object does not stand for the design itself but rather becomes a dialogue piece that aims to invoke curiosity and reflection. By raising critical questions,

163

the installation seeks to explore the socio-economic realities of international trade and provokes reflection and contemplation on the intricate and invisible threads that weave international black pepper tradein a postcolonial-capitalistic world.

Figure 9.3. Garag, Poorvi. Black pepper wallet. Fabulative product designed using critical fabulative design methodology. May 2024.

Reflection

This chapter contributes to the ever-growing international research on ethnographic design research and social design projects done in academic settings through the application of the ethnographic and critical fabulation methods and approaches. By examining the international trade of black pepper, the design project broadens the scope of design practices, incorporating global perspectives and embracing diverse, distinct experiences. The analysis of black pepper international trade is based on what I read and studied in school, my life experiences, the stories of colonialism I hear from many Indians, including those within my closest social circles. As such, subjectivity plays a pivotal role throughout the project's process.

The installation, which features a black pepper wallet and a historical timeline of black pepper trade, displayed at Dutch Design Week 2024. A recurring reaction from visitors was, "I didn't even know black pepper comes from India." This experience was echoed during its exhibition at the *Voedselverhalen Gouda Expositie* 2024–2025, which attracted attendees with a particular interest in food and history. People familiar with Dutch colonial history and its impact on the global spice trade found it refreshing to encounter an Indian perspective on a topic they had grown up studying. During my conversations with food historians and professors at the exhibition, the histories and narratives of India and the Netherlands were beautifully intertwined and brought to life within the exhibition room. This experience deepened my realisation that the seemingly mundane "table-top" spice is part of a wider constellation of social and material processes and structures, which makes the role of social design much more crucial to the world of design and the world at large.

The project "Peperduur" also travelled beyond the Netherlands, appearing in a design-focused exhibition in Prague. Visitors to the *Beyond (Design) Frontiers* exhibition 2024–2025, were amazed that pepper could contain such an intense and complex history to it. I saw design fulfil the role of storytelling and knowledge dissemination before me. While some engaged with the exhibition's design elements—the wallet, timeline, and video—others were drawn to the historical and narrative layers of the installation. The immediate connection between past and present was evident, as nearly every visitor had some personal experience with black pepper. The cognitive dissonance between a "precious black pepper wallet" and the everyday act of buying pepper for a few euros at the supermarket invited curiosity and reflective thought. I came to realise that through the installation and object design format, a slight shift in perspective and a slightly deeper understanding of design, black pepper, international trade and colonial history was possible. Visitors and the audience of the project were able to develop sensitivity to the subject of international trade and its colonial history. Developing an alternative narrative through critical fabulation allowed the audience and myself to engage in a world that could have existed,

165

enabling reflection on the world that currently surrounds us. The critically fabulated narrative could not have been developed without the ethnographic research done in phase one of the project, which evidenced that storytelling and creative exercises like this project only benefit from deeper research in contact with people—in this case, farmers and consumers of black pepper. Central to the philosophical approach of this project is the desire to uncover overlooked information and presenting it in engaging and reflexive ways. Here, design research transcends the act of creating new objects in an already maximalist world that is struggling with waste and instead becomes an inclusive exercise of questioning the present reality of society.

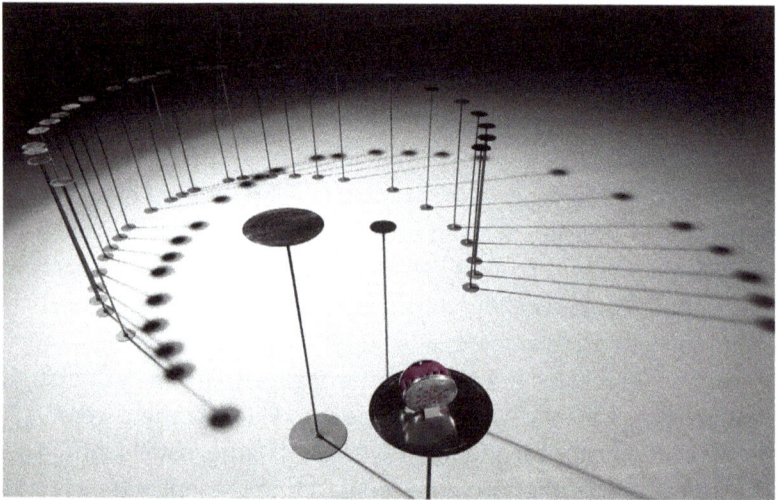

Figure 9.4. Garag, Poorvi. Exhibition design and 'black pepper wallet'. September 2024.

Key Takeaways

This chapter demonstrates how design research approaches like ethnographic research in combination with creative design methods like critical fabulation can effectively engage with complex societal topics. The key takeaways from this chapter are:

- This chapter establishes that design research approaches with novel methods can be utilised to dive into complex topics such that varied impacts, understanding and experiences can be elicited from the varied audience. The use of a literary method like critical fabulation for a design research and social design project enables design practices to go beyond design ideation and research methods and to take inspiration from related fields like anthropology, sociology, decolonial studies, future studies, literary methods, etc.
- Research is important to the creative process, and this is demonstrated in this chapter as without the extensive research in phase one of the project, the creative phase two would not have been achieved. By uncovering overlooked histories and subjective personal narratives—such as farmers and consumers in this case—enables a more nuanced and contextually rich design approach that is central to a creative social design project.
- The installation's success at Dutch Design Week, Voedselverhalen Gouda, and Beyond (Design) Frontiers builds confidence in the ability of design shows and installations to bridge historical narratives with contemporary experiences. By embedding history and trade within an everyday object like a pepper wallet, the project leveraged design to spark curiosity, cognitive dissonance, and reflective engagement, making complex histories and research more tangible, accessible and relatable.
- Through the experiences of the audience towards the critically fabulated narrative, the project also shows how design can provoke critical thinking and shift perspectives. Social design in this project, questions dominant narratives, engage audiences in reflexive ways, and contribute to deeper societal discourse—a shift from material production to knowledge creation.

167

References

• Albert Heijn. (2023). Drogheria zwarte peper. Retrieved November 1, 2023, from https://www.ah.nl/producten/product/wi65766/drogheria-zwarte-peper

• Appunn, K. (2021, March 5). EU's Farm to Fork strategy impacts climate, productivity, and trade. *Clean Energy Wire.* Retrieved from https://www.cleanenergywire.org/factsheets/eus-farm-fork-strategy-impacts-climate-productivity-and-trade

• Atkinson, P., Coffey, A., Delamont, S., et al. (2001). *Handbook of ethnography.* Sage Publications.

• Bhabha, H. K. (1994a). *The location of culture.* Routledge.

• Bhabha, H. K. (1994b). *The commitment to theory.* Routledge.

• Black Pepper Survey. (2023, December 20). Interview. Conducted by Poorvi Garag.

• Brazil Trade Business Group. (2023, August 3). The global pepper production. *Brazspice Spices - International Spice Brokers.* Retrieved from https://brazspice.com/market-news-offers/f/the-global-pepper-production

• Cartwright, M. (2021, June 9). The spice trade & the age of exploration. *World History Encyclopedia.* Retrieved from https://www.worldhistory.org/article/1777/the-spice-trade–the-age-of-exploration/

• Chen, D.-S., Cheng, L.-L., Hummels, C., & Koskinen, I. (2015). Social design: An introduction. *International Journal of Design, 10*(1), 1–5. Retrieved from https://www.ijdesign.org/index.php/IJDesign/article/viewFile/2622/719

• Cheyney, T. (2016). Historical materialism and alternative food: Alienation, division of labour, and the production of consumption. *Socialist Studies/Études Socialistes, 11*(1). Retrieved from https://socialiststudies.com/index.php/sss/article/view/24204/19975

• Dyck, J. (2019, October 7). When the Dutch ruled the world: Rise and fall of the Dutch East India Company. *Medium.*Retrieved from https://medium.com/bc-digest/when-the-dutch-ruled-the-world-rise-and-fall-of-the-dutch-east-india-company-57813dae4e72

• EIT Food. (2021, October 3). *The EIT Food Trust Report.* Retrieved from https://www.eitfood.eu/reports/trust-report-2021

• Garag, P. (2023). *Black Pepper Survey* [Unpublished survey].

• Hamer, C. (2020, July 28). Saidiya Hartman's critical fabulation can help inspire today's activists. *Study Breaks Magazine.* Retrieved from https://studybreaks.com/thoughts/critical-fabulation/?utm_content=cmp-true

• Haraway, D. (2016, May 24). Donna Haraway / Speculative Fabulation [Video]. *YouTube.* Retrieved from https://www.youtube.com/watch?v=zFGXTQnJETg

• Hartman, S. (2008). Venus in two acts. *Small Axe, 12*(2), 1–14. Retrieved from https://muse.jhu.edu/article/241115

• IDEO. (2019, July 9). What is social design? [Video]. *YouTube.* Retrieved from https://www.youtube.com/watch?v=QSs397-JF-Q

• Ivison, D. (2023, December 7). Postcolonialism. *Encyclopedia Britannica.* Retrieved September 3, 2023, from https://www.britannica.com/topic/postcolonialism

• Jackson, P. (2021). Food as a commodity, human right or common good. *Nature Food, 2*(3), 132–134. https://doi.org/10.1038/s43016-021-00208-z

• Kambar, C. (1994). *Oral tradition and Indian literature.* Sahitya Akademi. Retrieved from http://www.jstor.org/stable/44295584

• Kammar, U. (2023, December 20). Interview. Conducted by Poorvi Garag.

• Kearney, J. (2010). Food consumption trends and drivers. *Philosophical Transactions of the Royal Society B: Biological Sciences.* Retrieved from https://www.ncbi.nlm.nih.gov/pmc/articles/PMC2935122/

• Kneafsey, M., Cox, R., Holloway, L., Dowler, E., Venn, L., & Tuomainen, H. (2008, October). *Reconnecting consumers, producers and food: Exploring alternatives.* Bloomsbury Food Library. Retrieved from https://www.researchgate.net/publication/291941963_Reconnecting_Consumers_Producers_and_Food_Exploring_Alternatives

• Marxist Internet Archive. (n.d.). Karl Marx's theories on alienation. *Subject selection: Alienation.* Retrieved November 19, 2023, from https://www.marxists.org/subject/alienation/index.htm

169

- Mazlan, M. N., Saili, A. R., Saili, J., Zulkefli, F., Syahlan, S., & Ruslan, N. A. (2020, July 23). Exploring the challenges of pepper smallholder farmers in Sarawak: A qualitative study. *Food Research Journal.* Retrieved from https://www.myfoodresearch.com/uploads/8/4/8/5/84855864/_12__fr-mara-019_mazlan.pdf

- Nkrumah, K. (1965). *Neo-colonialism: The last stage of imperialism.* Thomas Nelson & Sons.

- Prashad, V. (2007). *The darker nations: A people's history of the third world.* The New Press.

- Said, E. (1978). *Orientalism.* Knopf Doubleday Publishing Group.

- Spivak, G. C. (1996). *The Spivak reader* (D. Landry & G. MacLean, Eds.). Routledge.

- Spivak, G. C. (2010). *Can the subaltern speak? Reflections on the history of an idea.* Columbia University Press.

- Subramanian, L. (2021, September 3). Coorg pepper. *Sāhasa.* Retrieved from https://sahasa.in/2021/09/13/coorg-pepper/

- Tsing, A. (2015). *The mushroom at the end of the world.* Princeton University Press.

- Verstegen. (n.d.). *Our history.* Retrieved November 1, 2023. https://algemeen.verstegen.nl/en/our-story/history-of-verstegen/

Peter Joore, Aranka Dijkstra

10. My Life as a Living Lab Coordinator

DOI: 10.1201/9781003609766

Introduction

This chapter is based on research conducted in collaboration with the AMS Institute for Advanced Metropolitan Solutions (AMS) in Amsterdam, The Netherlands. AMS was founded in 2013 as a cooperation between the municipality of Amsterdam, Delft University of Technology, Wageningen University, and MIT University, with the mission to design solutions for urban challenges and create better cities. An important aspect of AMS's approach involves the deployment of Urban Living Labs (ULLs). In these innovation environments, multiple stakeholders jointly create and test solutions to address complex metropolitan challenges. Examples are the Energy Lab in Amsterdam South-East, a neighbourhood with the ambition to become energy neutral in 2040. Or the Innovation Atelier in the Buiksloterham neighbourhood, aimed at developing Positive Energy Districts that create more energy than they use. Or the Marineterrein Living Lab, an urban setting in the core of Amsterdam where a variety of disruptive urban innovations can be tested before being replicated on the urban scale.

For the past years, AMS operated according to the Living Lab Way of Working (LLWOW) as described in Steen and Van Bueren (2017). An essential role in the deployment of ULLs is that of the Living Lab Coordinator, whose task it is to bring various stakeholders together, develop a shared vision, and facilitate co-creation and experimentation activities within the labs. This makes the coordinator an important audience for the LLWOW. In 2023, a review was conducted to assess the LLWOW based on the experiences accumulated during AMS's 10 years of existence. This analysis was conducted based on qualitative interviews and described in (Dijkstra & Joore, 2023). Based on the findings of this analysis, an updated Living Lab Way of Working has been developed, introducing a multilevel perspective as a basis for AMS's future way of working (Dijkstra & Joore, 2024). Key elements of the enhanced approach include (1) an emphasis on the iterative character of living lab processes, (2) the promotion of a shared vision the establishment of a common vocabulary among stakeholders, and (3) a holistic perspective regarding the necessary iterations between the macro-scale sustainability perspective on the one hand, and concrete experiments on the other.

In this chapter, we describe the added value of the LLWOW using a narrative. This narrative is a fictional combination of the experiences of different AMS Living Lab Coordinators. We use a narrative to communicate the developed insights because stories can help to contextualize the relatively abstract concepts we will discuss, making these insights more recognizable and accessible (Bruner, 1991). This is in line with the insight that the use of stories can help to increase reader understanding and engagement, allowing complicated information to be better processed (Green & Brock, 2000). Furthermore, stories help convey complex information in an accessible manner, helping to bridge the gap between experts and non-expert audiences (Dahlstrom, 2014).

So let's start! "Once upon a time, there was an enthusiastic professional who recently started working as a Living Lab Coordinator at the AMS Institute of Advanced Metropolitan Solutions..."

Once Upon a Time...

January 6: My New Life as a Living Lab Coordinator

"Today I started my new job as Living Lab Coordinator at the AMS Institute. Finally, a position where I can really work on my mission to help develop Amsterdam into the most sustainable, energy-efficient, clean, and intelligent city in the world! I am super excited to get started with this, although it's going to be quite a complex task. Even explaining what I will be doing turns out to be rather complicated. During Christmas dinner, I told my family that I will be working with Urban Living Labs, where stakeholders collaboratively test, develop, and create solutions to tackle complex metropolitan challenges. I noticed that they still found it quite challenging to fully understand. To be honest, I'm not entirely sure how everything

175

will unfold either, which is why I'm so excited to finally experience it in practice. I'm really looking forward to meeting all the people I'll be working with, and I'm sure they're just as enthusiastic about collaborating to shape the city of the future!"

January 10: Energy Positivity in Amsterdam

"So this was my first week at AMS. I survived! I am now officially introduced as the coordinator of one of Amsterdam's Urban Living Labs. The neighbourhood I work in wants to become 'energy positive'. This means that the area generates enough, or even more energy, than it uses itself. Ideally, the neighbourhood will become a kind of collective power plant. But before that can happen, several challenges need to be addressed, including the imbalance between energy demand and supply. During the day, energy is generated with solar panels. But often you need the energy precisely at night. The challenge is how to balance the two. That's quite a complicated challenge—I'll ask around to see if one of the university researchers might have a solution for this."

January 15: A Promising Solution

"Yesterday I explained our energy problem to one of the Delft researchers, and he came up with a promising solution. To better match supply and demand, he suggested that we can transform the energy that is generated during the day into hydrogen molecules. This hydrogen can be collected in an underground storage system. It then can be converted back into electricity at a suitable time (when there is little sun, but you do need the energy) using a technology called fuel cells. He explained that the technology is still in the process of development, though. But I suppose that doesn't matter. After all, we want to be an urban laboratory where new things are tried out, don't we? I am glad we have found a solution to the issue. Now all that remains is to bring the various stakeholders along with the idea, but I guess that can't be all that complicated."

January 21: Different Concepts and Definitions

"Today we had a meeting with the various Living Lab coordinators of the AMS Institute. I hoped that this would be a good opportunity to discuss a problem I was running into. During the last week I found out that there are quite a few different definitions being used, and I'm wondering what is the right one. Some people talk about Living Labs, or Urban Living Labs, while others call it a Fieldlab, a Testlab, or an Experimentation Environment. And yet others talk about a Community of Practice or a Hybrid Learning Community. That wouldn't be so bad, were it not that everyone is convinced that their definition is the only right one! Academics in particular are difficult to communicate with. They insist that you use the 'right' term, otherwise they refuse to talk to you any further. Yesterday a PhD researcher explained to me that our hydrogen initiative is not a real living lab at all, but that I should definitely call it an experiment. And then another PhD researcher presented me with the exact opposite perspective, just a few minutes later. To be honest I was somewhat happy to find out that my fellow coordinators also suffer from this problem. We had only limited time available, so we haven't discussed how to deal with it yet."

January 28: Different Roles and Approaches

"In our weekly meeting with the Living Lab coordinators, we discussed an interesting issue. As the concept of Living Labs may not be completely clear, it is even less clear what the job of a Living Lab Coordinator includes. To be honest, after our meeting, I was even further confused. Some of my colleagues indicate that they take the reins entirely in their own hands, taking the initiative and leading the way in setting all kinds of ambitious plans in motion: 'You have to show leadership, and get ahead of the troops, otherwise nothing will happen!', they explain. On the other hand, other colleagues stress that they absolutely take no initiative of their own, because all plans need to come from the residents and other stakeholders. 'I only facilitate the process and the conversation, bringing people together. Then the initiatives really have to come from the people themselves', they say. And still others explain—yes, well here it comes—that you have to do both at the same time, depending on the situation. So sometimes you have to leave it to the group, and sometimes you have to take the

177

initiative yourself. Situational coordinatorship is what they call it. Sounds good, but easier said than done. Soon we'll have a meeting where we start working on the hydrogen storage project. Then I can give my various coordination styles a try."

March 5: Exploding Hydrogen Plans

"Today I helped organise a meeting with all the neighbourhood residents and other stakeholders involved in the hydrogen energy storage initiative. I'm afraid that our promising and ambitious plans have been more or less derailed today, although the meeting started so positively. The idea is—or was—to build an underground hydrogen storage facility, where the energy generated during the day in the neighbourhood using solar panels is stored as hydrogen in large underground tanks. First, the chair of the local energy cooperative repeated the ambition to become a positive energy neighbourhood. Then, an enthusiastic technical engineer from Delft University presented the storage system and explained what it would take to install it. He was very positive about the possibilities and offered to help with the detailed design of the system. So far, everything was going according to plan.

"The second speaker was a local entrepreneur, who developed a comprehensive business plan for the initiative. It required a hefty sum as an investment, with amounts with quite a few zeros in it, but he expected that the initial investment would be recouped in five to seven years. And, if the government added some subsidy (it was, after all, an experiment aimed at the future sustainable city) the payback period could be even shorter. He introduced some enthusiastic investors who wanted to join in. Although the government official present looked a bit cautious when asked about available subsidies, he did indicate that he would certainly explore a possible contribution with the relevant department. So far so good. But then all hell broke loose.

"One resident had been reading up on the subject beforehand, and asked how risky such an underground hydrogen storage facility would be. The friendly engineer showed some slides with various risk profiles. I must admit they were quite complicated diagrams and tables, and the statistical calculations he presented were not really simple to understand either. But the probability of major explosions was virtually zero, he explained. To which the chairman of the neighbourhood association asked: 'Virtually zero, so that's not quite zero?' To which the engineer replied: 'No, the probability of risk is not 100% zero. Something can always happen. But statistically, that chance is extremely small.' The residents were certainly not satisfied with that answer, and immediately the room was filled with high-spirited discussions with statements like 'we don't want an underground exploding Hindenberg Zeppelin in our neighbourhood!' Residents then looked at the government official that was present, wondering whether the government would just allow its citizens to be exposed to this apparently rather risky experiment.

"It even got worse when the chair of the energy cooperative responded to all the commotion, stressing that we are all part of a local living lab after all. And wasn't this precisely created as a place to learn and experiment in together? Some residents turned out not even knowing that they were living in an Urban Living Lab. They had simply come to live here because it was such a nice house, and not to be part of a risky experimental neighbourhood. Well, I ask you!

"The worried government official grew increasingly cautious, indicating that things would not happen so fast. After all, no permits at all had yet been granted to get the initiative up and running. Indeed, the technology to be used was so new that it was not even clear yet what such a permit would have to comply with. And before that permit would be granted, a comprehensive risk analysis and environmental impact report would certainly be prepared. So that was certainly all going to take some time. At least a couple of years, he said.

"So, this comment again disappointed the entrepreneur who had spoken just before. He asked, 'What do you mean, that this is going to take some time? Just last week, your fellow government official explained to me that we are so progressive in Amsterdam, and that the city wants to lead the way in renewable energy initiatives. And now it turns out there are all procedures and protocols that work against it. What's more, those procedures apparently still need to be developed entirely. If this is going to take such a long time, then I better invest my money elsewhere!' He then walked out of the room stating that he wasted his time. Soon after, the government official also left, explaining that he had a very important meeting to attend. After which the chair of the energy cooperative thanked everyone for attending. And then it was over. Just like this one promising Living Lab initiative, in my opinion, is over too. To be honest, I think I'd better start looking for another job…"

June 12: A New Perspective

"I realise that in March I really did think my job was over. That was, until I shared my frustrations during our regular meeting with the other Living Lab coordinators. When I told them about the distressed citizens and the disillusioned entrepreneur, they didn't seem to be surprised at all: 'No worry, this is all part of it. But maybe the hydrogen storage project is a bit too much of a technology-push project. Was this really conceived by the stakeholders themselves?' I had to admit that the idea came mainly from the Delft engineer (who was looking for a suitable test environment for the new technology) and the inspired entrepreneur (who had a nice business plan in mind). To be honest, involving the citizens of the neighbourhood was something we had only scheduled at the end of the development process.

But what about the discovery that totally new policies and rules had yet to be developed by the government? This too turned out to be the case much more often. Some of my experienced colleagues pointed out that within these kinds of innovative developments, you have to switch between concrete experiments (such as the underground energy storage) and more long-term preconditions (such as the necessary regulations). It turned out that there is even a model to explain the relationship between them (see Figure 10.2). According to this model, the top level is called the 'innovation ecosystem level', and the bottom level is called the 'real-life environment level'. The Urban Living Lab is exactly the place to connect those two. At the end of the meeting, someone handed me a copy of the *Urban Living Lab Way of Working Handbook* (Dijkstra &

Joore, 2025), emphasising that I should definitely read it. Although I must confess that I'm not so much of a reader, as I prefer the action, I suppose that I do need some additional insights to save my project."

June 13: Eureka!

"After our meeting, I immediately started reading the *Urban Living Lab Way of Working Handbook*. I even tried to apply the multilevel model, based on my own experiences. Just filling it out made a lot of things clear. Some people operate mainly at the lower level, preferring to work on practical local experiments, while others prefer to focus on the bigger picture and aim to understand the overall innovation ecosystem. Both perspectives are essential to achieving successful change. There are also several arrows between the different layers. For instance, a downward arrow visualises the identification of a problem in the area of energy balance. This results in a need or opportunity, such as the requirement for an innovative way to store energy.

"The upward arrow refers to the implementation of a new innovation and its integration into the broader system. Regarding the idea of hydrogen storage, the technology itself appeared to work (according to the engineer), but its integration into the larger system proved challenging. This was partly because the residents were not particularly enthusiastic about the new system (to say the least) and because no regulations were in place to properly manage the process. This illustrates how the different levels influence one another, both from top to bottom (the need for a new energy storage system) and from bottom to top (the necessity for new regulations and the acceptance, or lack thereof, of a specific technology). Everything is connected to everything, as it were. Surely it would be great if my new insights could help take our Living Lab one step further. I'm going to think about how we can get that done."

181

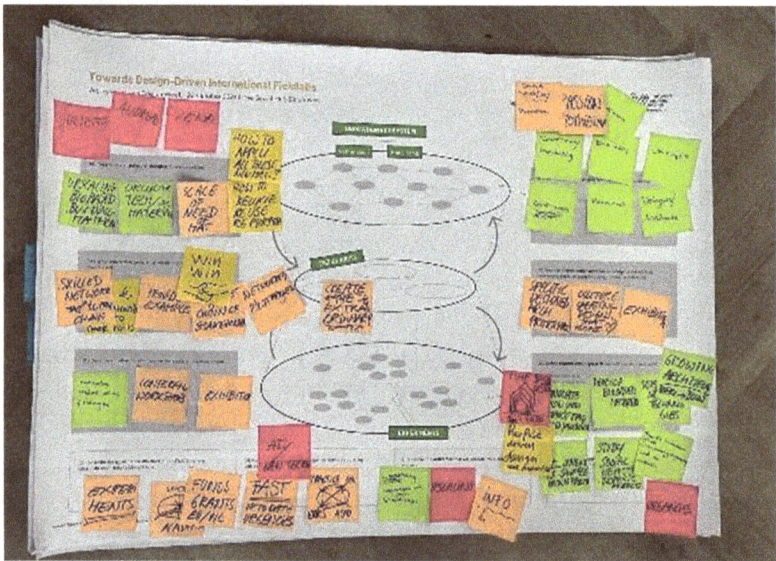

Figure 10.1. Work session based on the LLWOW model (top) and the filled LLWOW model (bottom). Photos taken during Dutch Design Week 2024, at a work session based on the ULLWOW.

Collaborative Platform

An organizational structure that connects stakeholders across disciplines, enabling knowledge sharing, experimentation, collective learning, and scaling of urban innovations.

Innovation Ecosystem

A network of interconnected stakeholders who influence and contribute to urban transition processes through their individual and collective actions.

Stakeholders

Individuals, groups, organizations, policymakers, businesses, researchers, and citizens involved in or affected by an urban transition process.

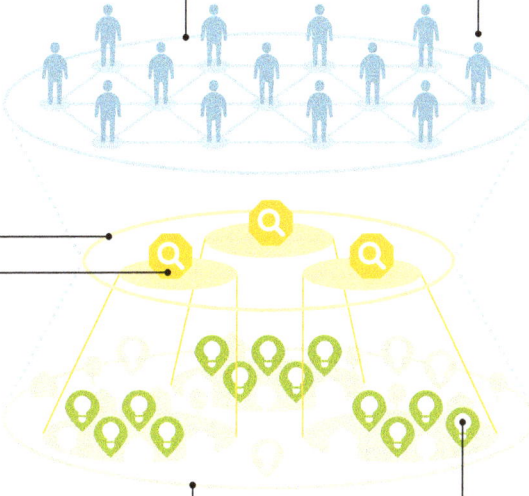

Urban Living Labs

Collaborative settings where local stakeholders co-create, test and evaluate innovative solutions in real-life environments, to address complex urban challenges, with the goal of scaling or replicating them across the city.

Real-Life Environment

A physical or virtual environment (e.g. a street, building, or digital platform) where innovative ideas and solutions are tested and developed under real-world urban conditions.

Experiment

A structured activity designed to develop and validate new ideas, accelerating learning and generating insights through real-world testing and experience-based learning.

Figure 10.2. Urban Living Lab Multilevel Framework (from Dijkstra & Joore, 2025, page 31).

October 23: Good News, and Bad News

"Good news, our Living Lab is still (or again) alive and kicking! Inspired by my colleague's suggestions, and the new systemic perspective that I discovered in the Urban Living Lab Way of Working Handbook, I mustered up the courage to get back into the conversation with the various stakeholders in our neighbourhood. To cut a long story short, it worked out rather well. Based on several collaborative working sessions, where we started thinking again from scratch, it turned out that local energy storage was still essential to realise the shared ambition ('we want to become a positive energy neighbourhood').

183

"However, the underground hydrogen storage proved a step too far. The new technology appeared to be too much in its infancy, with many technical uncertainties that one would preferably explore at a safe distance from inhabited areas. Fortunately, the engineers did find another experimental location, somewhere in the Groningen countryside. There, they can experiment with their new storage technology at a considerable distance from civilisation. I understand there is even a high fence around their test location, so no one can really get near it and get hurt. The unfamiliarity with the new technology, in turn, created uncertainty among the neighbourhood's residents. Naturally, they didn't want to be guinea pigs, especially considering the potential risks involved. And as I already mentioned, the necessary regulations still needed to be developed. In short, the plan for hydrogen storage was shelved. That's the bad news. The good news is, that next week we'll have a workshop discussion other options and alternatives."

October 30: Welcome to Community Batteries

"Today we had a productive working session, in which a group of determined participants started looking for a solution to store energy that was slightly less in its infancy. After exploring various options, we ended up with the concept of a community battery. The technology for this is much more mature, and residents are well acquainted with batteries from their own experience. As a result, concerns about disasters and problems are much less prevalent. I am really looking forward to developing this plan further."

November 27: The Sea Containers Are Coming

"It was still a puzzle to develop a suitable business model, but with energy prices fluctuating wildly, the concept of a community battery is proving to be very suitable for storing energy when energy prices are low, and using the energy again when energy prices are high. Work is now underway to install three big sea containers somewhere next year. One container will contain a large lithium-ion battery, the second an electrical power converter, and the third container will be set up as a visitor centre to share knowledge about the battery. This way, we are also doing communication and awareness-raising. I love it when a plan comes together!"

December 23: Looking Back on My First Year

"Looking back on my first year as Living Lab coordinator, I learned quite a lot. When I first started, I wondered what exactly my job entailed. Now I understand that I am actually a kind of connector. A connector of people and organisations. And a connector of developments taking place at different system levels, linking big visions and ambitions with small and concrete actions. I realise that, as a Living Lab coordinator, I am not a resident of the neighbourhood myself, nor am I an entrepreneur who works there. But connecting residents to entrepreneurs, government, and knowledge institutions like universities, I do contribute to positive developments in the neighbourhood.

"By the way, when I just started working here, I was rather critical of those researchers who mainly put things on paper without actually doing anything themselves. But I must confess that the insights described in the Urban Living Lab Way of Working have really helped me to do my job. First, it helped to create a similar concept of an Urban Living Lab, by offering a common vocabulary. Second, it inspired me to better distinguish between the various perspectives and interests of different stakeholders. Coordinating a Living Lab really requires many different skills and actions, and the multilevel model really supported me in positioning all those activities in relation to each other. Maybe I will write a book myself one day: 'How I survived as a Living Lab Coordinator."

But first, it's time to take a Christmas break. An excellent time to reflect on the lessons I have learnt this year. I think my most important lesson is that true transformation does not happen overnight. Developing an energy-efficient, clean, and intelligent city isn't done in one day. It is the small, steady steps forward that together add up to meaningful and lasting change!

185

Key Takeaways

Based on the above fictional story of a Living Lab coordinator, we propose the following takeaways for practitioners:

- When starting a project involving different stakeholders, discuss and agree on the various concepts and definitions, and document these in a shared vocabulary. If new partners join the project, this vocabulary can help them quickly align with the ongoing discourse. This prevents repeated discussions about definitions. The Urban Living Lab Way of Working (Dijkstra & Joore, 2025) could serve as an initial framework for such a vocabulary.
- Involving various stakeholders is a crucial aspect of a Living Lab. The composition of the stakeholder group may change throughout the duration and different phases of the Living Lab. Therefore, it is important to effectively facilitate the process of stakeholders joining and leaving during the lifespan of the Living Lab.
- Developments at micro and macro levels mutually influence each other, in both top-down and bottom-up directions. Ensure regular iterations where developments taking place at a high systemic level are aligned with concrete innovations at a more practical level.
- Whether an Urban Living Lab is the most appropriate means to investigate a particular innovation partly depends on the development status of the technologies being deployed. If a technology is not yet sufficiently developed, applying it in a Living Lab may be too premature. If the technology is still in an early development phase, testing in a different type of laboratory setting may be more appropriate.

References

• Bruner, J. (1991). The narrative construction of reality. *Critical Inquiry, 18*(1), 1–21. https://doi.org/10.1086/448619

• Dijkstra, A. M., & Joore, J. P. (2024). *Towards a Living Lab Way of Working 2.0.* Amsterdam Institute for Advanced Metropolitan Solutions.

• Dijkstra, A. M., & Joore, J. P. (2025, forthcoming). *The Urban Living Lab Way of Working.*.Amsterdam Institute for Advanced Metropolitan Solutions.

• Dahlstrom, M. F. (2014). Using narratives and storytelling to communicate science with nonexpert audiences. *Proceedings of the National Academy of Sciences, 111*(Supplement_4), 13614–13620. https://doi.org/10.1073/pnas.1320645111

• Green, M. C., & Brock, T. C. (2000). The role of transportation in the persuasiveness of public narratives. *Journal of Personality and Social Psychology, 79*(5), 701–721. https://doi.org/10.1037/0022-3514.79.5.701

• Steen, K., & van Bueren, E. (2017). *Urban Living Labs: A Living Lab Way of Working.* Amsterdam Institute for Advanced Metropolitan Solutions (AMS).

Part 3:
Balancing Worldviews

"If applied design research is to have a profound impact on societal transitions, a delicate orchestration of diverse worldviews, priorities, and methodological approaches is a necessary precondition."

Bas de Boer, Roelof A.J. de Vries, Mailin Lemke

11. Research Culture Clash in Behaviour Change Design: Three Dilemmas for Behaviour Change

DOI: 10.1201/9781003609766

Introduction

Design undoubtedly has a strong influence on the way we live and act. This influence manifests both on a small scale, through the creation of artefacts that are used in people's home environment, and on a larger scale, when designing buildings and urban infrastructures (Niederrer, Clune, and Ludden, 2018, 3). A relatively new field of design focuses specifically on designing for behaviour change and can be termed "behaviour change design" (BCD). The ideal outcome of BCD is the realisation of positive behaviour desired by both society and individual users. Understanding and analysing the emergence of particular behaviours is key to BCD, and this is done by, amongst other things, user psychology and behavioural theories and models (Khadilkar and Cash, 2020, 517–518; Niederrer, Clune and Ludden, 2018). As a result, BCD has a highly interdisciplinary character. In this chapter, we provide a critical analysis of research in the context of behaviour change and design applications between design-focused disciplines (e.g., human-computer interaction (HCI) or interaction design) and behavioural scientists, specifically focusing on the problems emerging due to the interdisciplinary character of BCD.

In an academic context, interdisciplinary efforts are often thought of as the only way to tackle real-world problems that are complex and not confined to the boundaries of individual academic disciplines. However, it seems that the skills and competencies of the different researchers involved often prove to be more challenging to integrate and synthesise than envisioned on paper (Fischer, Tobi, and Ronteltap, 2011, 348). Some of the barriers that might hamper interdisciplinary research can be traced back and are likely reinforced by academic and funding structures. These can give rise to institutional barriers that can play a role in career advancements in academic contexts (e.g., Jacobs and Frickel, 2009, 47). Other barriers result from the difference between design research (and academic research more generally) and design practice because designer practitioners have to take into account values that differ from those of academic research, such as design briefs formulated by clients, as well as details of commercialisation, reliability, costs and profitability of design concepts (Kolko, 2010, 80; Norman, 2010, 9-10).

In this chapter, we focus on another type of barrier, namely one originating in the differing research cultures across research disciplines *within* an academic context, and explicitly focus on differences between design research and behavioural science. Since researchers are trained within a specific discipline, they are familiarised with certain research traditions, acquire specific theoretical beliefs, inherit a particular perspective on the extent to which their research should be practically applicable, and generally rely on implicit assumptions about what constitutes good research. We focus on the differences in the research cultures of design research and behavioural science and suggest that differences in these two research cultures give rise to three dilemmas in BCD: (1) the perception of the degree of complexity of a problem, (2) the ideal process to address the problem, and (3) the overall goal of assessing the influence of the BCD on the user. Formulating these dilemmas is an attempt to make the presuppositions of the different research cultures of design and behavioural science explicit.

The chapter is structured as follows: First, we briefly discuss why interdisciplinary research is an ideal in BCD. Second, we clarify the notion of research culture and show how it can be used to analyse how (often implicit) scientific commitments and beliefs structure what constitutes good research. Third, we introduce three dilemmas that researchers focusing on interdisciplinary BCD face. In conclusion, we argue that the awareness of the existence of such dilemmas and the ability to explicitly position oneself in relation to those contributes to the realisation of interdisciplinary works in practice.

BCD: An Interdisciplinary Ideal

Behaviour change interventions are often delivered as part of complex systems consisting of specific, fine-grained, and replicable components referred to as "behaviour change techniques" that aim to redirect or alter causal processes influencing one's behaviour (Michie, van Stralen, and West, 2011, 8). Psychological theories play an essential role in designing behaviour change interventions, as these help in explaining when, how, and why specific changes may occur (Michie and Johnston, 2012, 4).

While changing user behaviour in an "ideal" direction may seem straightforward in theory, such attempts are often complex. What complicates the development of concrete BCDs (e.g., using specific designs to increase physical activity level among school children) used as part of a larger behaviour change intervention even further is that theoretical components are challenging to translate into concrete designs to influence a specific behaviour (Cash et al., 2022, 10-11, see Figure 11.1). In addition, aspects such as aesthetic, moral, and contextual appropriateness might play a role in how the BCD is perceived (Van Arkel and Tromp, 2022).

Figure 11.1. Different considerations when creating specific BCD, e.g. to support school children be more active. Copyright © 2024 Authors.

It has been argued that to facilitate this process and develop efficient and effective BCDs, research should be conducted by interdisciplinary teams. Furthermore, it has been argued that the combination of design research and psychology is especially suitable for BCD. Such interdisciplinary teams allegedly combine the best of both worlds by merging designers' creative, imaginative work with the systematic research done in psychology (Hallsworth and Kirkman, 2020; Khadilkar and Cash, 2020, 524; Reid and Schmidt, 2018). However, it seems often more difficult to integrate the insights of different disciplines in practice than envisioned on paper. In the next paragraph, we draw from sociological and philosophical studies to point to potential barriers to interdisciplinarity.

Differing Research Cultures as Barriers to Interdisciplinarity

Sociological or policy studies on interdisciplinarity tend to emphasise the *institutional* barriers that might hamper interdisciplinary work. These include universities being organised around specific disciplinary lines, the absence of interdisciplinary teaching in educational programs, or that most grants and funding opportunities are directed towards specific disciplines (Huutoniemi, 2010, 313). The development of new funding schemes organised around specific societal challenges, the attempt to reform educational programs, and the explicit promotion (and rewarding) of collaborations with researchers in different fields, are put forward as effective ways of contributing to an interdisciplinary environment (Jacobs and Frickel, 2009, 48; Pellmar and Eisenberg, 2000, 5–6).

However, while such changes definitely promote a more interdisciplinary environment, a closer look at concrete scientific practices reveals that researchers continue struggling to transcend disciplinary boundaries and cooperate successfully (MacLeod, 2018, 716–717; Reay, Craig, and Kayes, 2019, 394–398). Also, applied interdisciplinary work in the context of HCI, design, and

195

healthcare indicates that this approach can be a challenge due to different research approaches, goals, expectations, and regulatory frameworks (Blanford et al., 2018; Groeneveld et al., 2018; Moody, 2015, 400).

In this chapter, we propose to understand problems emerging within such interdisciplinary collaborations as resulting from *research culture barriers*. These barriers emerge from the different background concepts, methods, tacit knowledge, technologies, and epistemic standards of different disciplines (Jacobs and Frickel, 2009, 47; Keestra, 2017). These differences, we argue, undermine the possibility of the establishment of mutual understanding between disciplines, which is needed for a successful convergence eventually leading to the development of a solution to an interdisciplinary problem.

The idea that scientific work requires specific conceptual, epistemological, and methodological frameworks, and that these frameworks differ across disciplines can be traced back to Thomas Kuhn's notion of the *disciplinary matrix*. According to Kuhn, normal science proceeds in a certain scientific community that is organized around certain shared theoretical beliefs, values, scientific instruments, experimental techniques, and even metaphysical beliefs (Kuhn, 2012, 181). This means that there (most often) is a disciplinary consensus when it comes to which problems (or puzzles as Kuhn calls them) should be central to scientific work, how these problems should be addressed, and when a problem can be said to be solved (Andersen, 2016, 7). Scientific practitioners themselves are not necessarily consciously aware of their own disciplinary matrix, nor are they always capable of articulating the specific commitments that structure their research. Given that such matrices are *disciplinary*, standards tend to differ across disciplines. And since they are derived from a background that is not easily made explicit, scientists might experience problems when communicating or having a dialogue about their epistemic standards.

Whereas Kuhn's ideas derive from an analysis of case studies in the history of science, recent sociological and ethnographic studies confirm that scientists continue to have strong disciplinary commitments that are explanatory of the success of individual disciplines. For instance, Karina Knorr-Cetina has closely examined how particle physicists and molecular biologists each have very different practices through which they construct their objects of knowledge and interpret them. She argues that these different disciplines embody different *epistemic cultures* that subscribe to different standards of knowledge production (Knorr-Cetina, 1999, 8–9). Other studies have revealed similar issues: the different research traditions of experimental and computational biologists make it difficult to construct shared research goals, ecologists and economists adhere to different standards when it comes to modelling (MacLeod, 2018, 708), and the integration of brain science and psychology proves difficult when studying human cognition (de Boer, 2021, 190; de Boer, te Molder, and Verbeek, 2020, 519).

Given that there are many barriers to interdisciplinary collaborations within the same general field (e.g., molecular biology and computational biology), or between neighbouring fields (e.g., brain science and psychology), it should not come as a surprise that similar challenges occur in the context of BCD. However, surprisingly little research on challenges and pitfalls in interdisciplinary collaboration between design researchers and behavioural scientists has been carried out thus far (Khadilkar and Cash, 2020).

In this chapter, we intend to offer a starting point for understanding the barriers to interdisciplinary collaboration between behavioural scientists and designers in terms of their research cultures, instead of one that relies on an a priori distinction between scientific and design methods. Having scientists and designers inhabit different research cultures with different epistemic commitments allows development of a nuanced starting point for understanding the emergence of barriers to interdisciplinary research.

197

Three Dilemmas

We take the different ideal research processes in design vis-á-vis behavioural science as depicted in Figure 11.2 as a point of departure for the formulation of the three dilemmas. Both depicted research processes are idealizations. Design researchers are likely to have multiple iterations and move back to a previous step to start the research process again. The ideal of behavioural science is more linear and reflects how articles are typically structured and how research is presented, whereas the practice of doing research is vastly more complex and messier (e.g., Latour, 1987, 5). The different dilemmas are illustrated through the example of BCD for health.

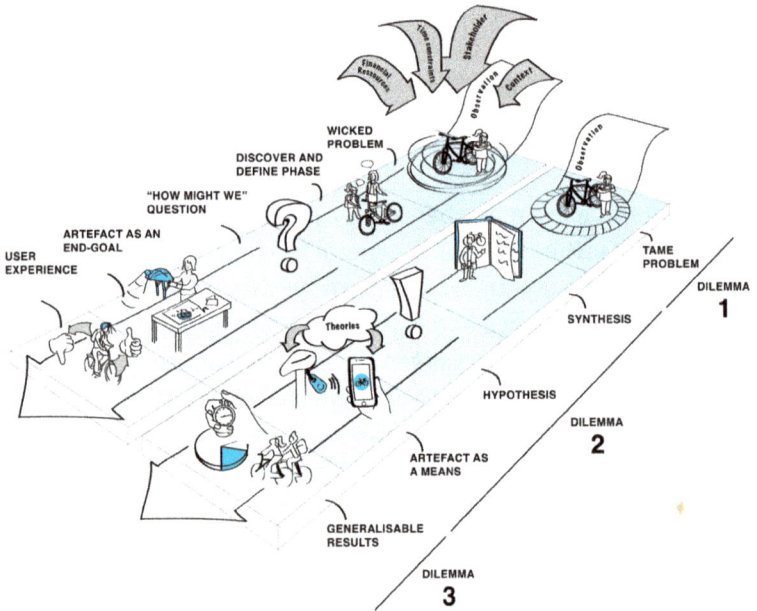

Figure 11.2. Different dilemmas encountered in interdisciplinary research. Copyright © 2022 Authors.

Dilemma 1: What's the Problem?

The first dilemma involves how the problem to be addressed by a BCD should be defined. Take, for instance, health promotion. Health behaviour is complex in that the behaviour of individuals is shaped by physical, psychological, economic, socio-cultural and

198

political factors (Signal et al., 2013). As a result, it is often unclear at what level to intervene, or how an intervention on one level affects and/or is affected by another level. Furthermore, different stakeholders are involved, each having a different view of what counts as healthy behaviour and what kinds of interventions are desirable or possible. Due to this complexity, it is difficult to precisely define the problem on which BCD should focus.

For this reason, health, or health promotion, is sometimes referred to as a so-called *wicked problem* (e.g., Signal et al., 2013). This term can be traced back to the work of Horst Rittel, who suggested that there are many problems that are, at their core, ill-formulated, confusing and influenced by different stakeholders with often conflicting values (Churchmann, 1967, B141). In a later paper, Rittel and Webber outline in more detail how the projects that are carried out by planning and governing systems appear to be inherently different from the problems that classical science focuses on (Rittel and Webber, 1973). They make a distinction between *tame* problems that are addressed within the classical sciences, and *wicked* problems that present themselves in planning and governance. The former category comprises a set of problems that are—or can be—well-defined: all relevant elements can be determined and all requirements can be specified, such that a linear step-by-step model towards a solution can be applied. The latter category, by contrast, lacks this possibility of clear specification.

Building on Rittel's work, it has been suggested that tame problems are at their core "tamed" wicked problems where researchers try to address a smaller part and are essentially "carving off" a piece of the problem and finding a rational and feasible solution to this piece" (Churchman, 1967, B141). Therefore, every tamed problem necessarily *misses out* on the complexity of the wicked problems, and the strength of design research lies in the ability to preserve the "wickedness" of the problem at hand. On the other hand, behavioural scientists tend to celebrate their capacity to cut larger problems into smaller and measurable pieces to yield robust solutions. From this point of view, wicked problems are only *apparently* wicked, because they are not well-formulated *yet* and tame problems are "diminished versions" of "wickedness" (Coyne, 2005, 8). Following this line of argument, the solutions to a set of well-formulated (tame) problems will eventually converge into a solution to the larger (wicked) problem to which they are connected. Such a perspective brings us back to the first dilemma occurring in multidisciplinary work between designers and behavioural scientists: should wicked problems be tamed, or should the wickedness of problems be cultivated?

Case Study: Explorative Self-experimentation

We refer to a previous study by one of the authors to illustrate this dilemma. The study by Fedlmeier et al. (2022) aimed to support people in changing health-related behaviour. This goal was perceived as a wicked problem from the start and was based on the belief that there is no "one-size-fits-all" solution or behaviour change technique when it comes to changing a specific behaviour. Based on an explorative and iterative design process, the study suggests several factors that designers should consider when creating BCD, see Figure 11.3.

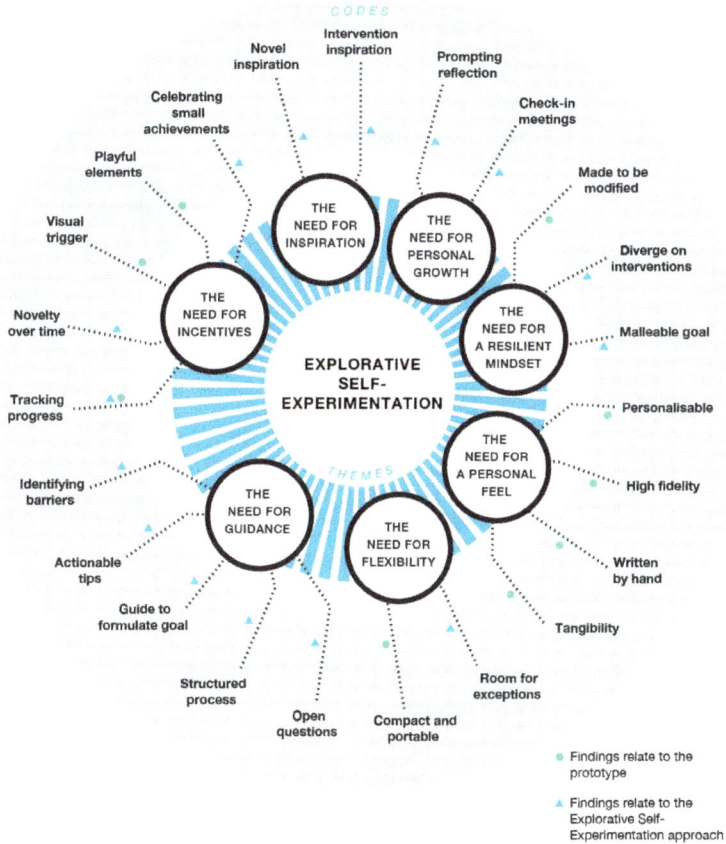

Figure 11.3. Overview of aspects designers can consider when addressing the wicked problem of supporting behaviour change through self-experimentation. Copyright © 2022 Fedlmeier et al.

However, from a tame problem perspective, the same goal of supporting people in changing a behaviour is likely to skip such an explorative research approach and focus on one specific and predefined barrier or facilitator to create a BCD. Contextual factors and requirements that go beyond those of the target group might play no role in the perception of the problem.

Dilemma 2: What Is the Ideal Research Process?

A second dilemma emerges when considering the different research processes by which a problem is approached. Depending on the research culture, different steps are followed to address the problem and develop a particular design. The differences between the research processes in design research and behavioural science can be clarified by pointing to the different forms of accountability to which they typically need to conform. The term accountability refers to "the expectations of what activities must be defended and how, and by extension the ways narratives (accounts) are legitimately formed about each endeavour" (Gaver, 2014, 147). In line with Gaver (Ibid., 147–148), we suggest that the research steps in design take place against a background of *aesthetic accountability*, whereas the research in behavioural science can better be understood in terms of *epistemological accountability*.

The classic research process used in behavioural science follows a version of the scientific method (see Figure 11.2 for a simplified version). This process aims to make it possible to be held epistemologically accountable for the research process. From this perspective, researchers should ideally first explore the literature to synthesise findings from related work and base studies on a theoretical framework. Based on this process, one or multiple *hypotheses* are formulated. Conditions for formulating a hypothesis are the restriction to a specific research domain and the development of clear and measurable variables.

On the other hand, the design research process focuses on aesthetic accountability. "Aesthetic" here does not refer to how beautiful the design might be, but instead to the satisfactory integration of multiple design features (Ibid., 147). The process followed can be exemplified with the design thinking process (e.g., the double diamond) consisting of discover, define, develop, and deliver phases (Ball, 2019). The first step focuses on gaining insights into the problem (discover phase) and then scoping down the project (define phase). In this case, the purpose of these two phases is not to develop a well-formulated hypothesis but to create a question tailored explicitly to the pragmatic context in which the design should function. This perspective implies being specific about the concerns and needs of the different stakeholders that will be affected by the BCD. This process results in the formulation of a *"How might we?"* question guiding the design process, e.g. see Figure 11.4 for an example.

Figure 11.4. Example for a "how might we" question as a starting point for the design process. Copyright © authors 2024.

The second way in which the distinction between aesthetic and epistemological accountability manifests is the creation of the design that aims to evoke the ideal behaviour change. From the perspective of aesthetic accountability, the creation and final form of the artefact itself can be a *focus of attention* during the research process. From the perspective of epistemological accountability, the artefact is instead a *means* towards something else (e.g., knowledge generation through hypothesis testing).

In the context of design, the development of the artefact and its inherent qualities can be central to the research process (Dalsgaard, 2017; Zimmerman, Forlizzi, and Evenson, 2007, 496). The process of doing design is perceived as doing research and producing relevant knowledge (Stappers and Giaccardi, 2017). The seemingly "drifting" in design experiments can be perceived by behavioural scientists as uncontrolled, inconsistent and illogical but such experimental activities have been argued to facilitate knowledge building and leading to high-quality (and novel) work (Krogh, Markussen, and Bang, 2015, 42). The question of accountability in this context is aesthetic in that it focuses on *if* it works rather than outlining how the researcher knows this to be true by unravelling the underlying mechanisms (Gaver, 2014, 147–148). A factor that can influence the aspect of aesthetic accountability is if the developed design is "unique" and therefore demonstrates a significant invention rather than an improved copy of a previously introduced design (Zimmerman, Forlizzi, and Evenson, 2007, 499). This is likely to encourage designers to hold each other accountable in terms of innovativeness and creativity, which are criteria that are likely to conflict with the ones present in the research culture that behavioural scientists inhabit.

Case Study: Sensory Interactive Table

We refer to a study by one of the authors to illustrate this dilemma. The study by de Vries et al. (2020) explored through a research-through-design process how social eating could be supported by an interactive table. The authors conduct different activities and illustrate the social aspect of eating through three scenarios to explain the design rationales for form, materiality and modularity leading to the form and functionality of the Sensory Interactive Table (SIT).

Figure 11.5. Three scenarios for using the Sensory Interactive Table. Copyright © 2020 de Vries et al.

From the point of view of *epistemological accountability*, a focus would be on creating a hypothesis rather than explaining the reasoning of the design decisions in detail. For example, a hypothesis could be 'a round table will increase social interaction among diners'. What exactly this round table would look like, and which process was followed to develop the form, would not play a crucial role in this context.

Dilemma 3: What's the Goal?

The third dilemma focuses on the perception of the goal of the study. Behavioural science typically focuses on the development of generalizable results, whereas design research tends to focus on gaining detailed and specific insights into the user experience when evaluating the effect of a BCD. These two ends of the spectrum often have different standards of the ideal methods used to collect and analyse the data and sample sizes. For example, formative usability evaluation commonly focuses on determining early usability issues, refining the design, and often employing smaller samples and qualitative methods (Hartson, Andre, and Williges, 2003, 149–150). This type of testing aims to learn how specific users experienced the designed object/intervention, whether or not the design fits their particular needs, and what kind of conceptual or design-specific elements pose a usability issue, see Figure 11.5. Such an approach minimises costs and allows the development of new iterations of the concept quickly. However, it also limits the possibility of running long-term and extensive user studies as the design prototype is likely to include some technical, material or software shortcomings. In addition, point or case studies that are often used in design projects to evaluate a design prototype allow just very constrained speculations about the design and characteristics influencing the interaction as a baseline condition usually is missing. This and a focus on *if* a design works rather than *why* it works limits insights into features evoking a specific behaviour change in users (Oulasvirta and Hornbæk, 2022, 147–148).

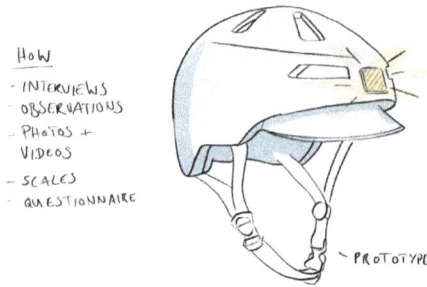

How
- INTERVIEWS
- OBSERVATIONS
- PHOTOS +
  VIDEOS
- SCALES
- QUESTIONNAIRE

PROTOTYPE

Figure 11.6. To evaluate a design prototype to make children feel safe when biking at night, formative usability testing can help to find points of improvement and evaluate if user needs are met. Copyright © 2024 authors.

From the perspective of generating generalizable results, the focus is on collecting quantifiable user insights involving large samples that allow the study results to be generalizable to the population level. Artefacts evaluated from this perspective aim to assess how the operationalization of factor X influences population Y at large. In the context of health, this is often done using randomised controlled trials (RCTs) to prove the effectiveness of an intervention (Blanford et al., 2018, 4). RCTs are a gold standard in medical research and aim to reduce any kind of bias by randomly allocating participants to treatment groups to evaluate the effectiveness of a new treatment or intervention (Hariton and Locascio, 2018, 1716). From this perspective, behaviour change standards should attempt to adhere to standards similar to the ones in medical research. When aiming to produce generalizable results, individual user experience plays a minor role. The sample size, allocation of participants to the three conditions, statistical and research methods, and reliability all play a crucial role in this context. The collected data is then compared to evaluate the intended effect. Following this approach, the findings can be generalised.

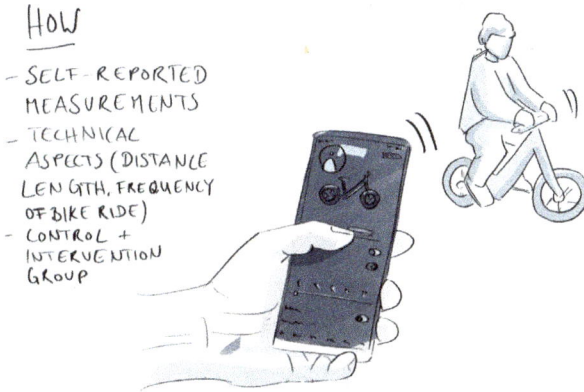

Figure 11.7. Generalisable results help evaluate the effectiveness of an intervention, e.g. a bike app with a personalised avatar to increase physical activity among children. Copyright © 2024 authors.

Case Study: The Biggest Hit

An example for this dilemma can be seen in the case study by Lemke et al. (2017). The case study explores the possibility of creating an everyday object to encourage people with stroke impairments of the upper limb to overcome so-called learned nonuse and encourage the use of the affected arm and hand. The developed prototype was tested with experts to develop an improved version of the prototype. Due to the small sample size, evaluation methods and testing length, results cannot be generalised. In the context of health and behaviour change, such findings are crucial to illustrate the potential of devices to health practitioners and patients.

Figure 11.8. The prototype to encourage a behaviour change was tested by stroke therapists due to ethical constraints. Copyright © 2017 Lemke et al.

The *concrete challenge* in the context of interdisciplinary work in BCD is how the often qualitative investigations into user experiences and the generalizable results obtained by quantitative studies can mutually inform one another and can culminate into a mixed methods approach. A starting-point for addressing this challenge and to find ways to successfully engage in interdisciplinary research requires acknowledgement that these different ways of doing research are the result of inhabiting different result cultures, instead of rendering one way of doing research epistemologically inferior.

Conclusion

The purpose of this chapter was to highlight some of the potential barriers to interdisciplinary research in behaviour change. Such interdisciplinary collaborations are often framed as highly desirable, as the only proper response to the complex nature of social problems, or even as an accomplishment on their own. This enthusiasm has resulted in the emergence of many initiatives on an institutional level to foster interdisciplinary research and to encourage researchers to move beyond their disciplinary boundaries. By specifically focusing on collaborations between designers and behavioural scientists in the field of BCD, we intended to highlight that specific barriers remain present in interdisciplinary scientific practices despite the relative success of institutional initiatives in putting scientists from different disciplines in the same room.

The three dilemmas that we identified show how difficulties arise in interdisciplinary collaborations—despite good intentions—as a result of researchers from different disciplines inhabiting different research cultures. Different research cultures adhere to different epistemological standards, use different methodologies and instruments, and might even have different metaphysical beliefs. Even more, it has been repeatedly pointed out that a research culture forms a background that is difficult to make explicit, making it challenging for scientists to communicate or have a dialogue about their own presuppositions about what constitutes good research. We take it that the two above-mentioned aspects indicate that (a) an exclusive focus on the institutionalisation of interdisciplinary research misses out on several important barriers to successful collaborations in practice, and (b) that insofar as research cultures constitute implicit barriers to interdisciplinary research, the making explicit of these barriers might be beneficial for researchers in an interdisciplinary setting. This might be specifically beneficial when setting up new interdisciplinary projects.

When understanding the difference(s) between design and behavioural science as stemming from different research cultures, instead of assuming an (almost) unbridgeable gap between them, a more helpful dialogue between the different disciplines can be established. Of course, the field of BCD goes beyond design research and behavioural science, and further work is needed to get an understanding of additional gaps that can be attributed to differing research cultures. In addition, our focus in this chapter is on design researchers rather than design practitioners who create artefacts likely to have a more significant effect on people's lives than those created as part of research projects. The research-practice gap has been pointed out in the context of design practice and design research before. This will require further attention to determine dilemmas specific to this translational science area where behaviour-change-specific scientific knowledge is translated into practice and back (Colusso et al., 2019; Khadilkar and Cash, 2020, 519–520; Norman, 2010).

Key Takeaways

Removing institutional barriers to interdisciplinarity is an important step forward but is unlikely to contribute to good interdisciplinary research on its own. This is because researchers from different disciplines inhabit different *research cultures*. The dilemmas that we introduced can be a helpful starting point for removing barriers caused by inhabiting different research cultures because they help establish a dialogue about three fundamental questions in the research process: 1) what is the problem that needs to be addressed, 2) which approach can be used to tackle this problem, and 3) what the end-goal of the research is. To facilitate an interdisciplinary discussion towards this end, our call to design researchers as well as behavioural scientists would be to explicitly position themselves in relation to these questions or dilemmas, and do the following:

- Consider the complexity of the problem at hand, e.g., is it wicked or is it tame?
- Consider the type of accountability that is aimed for in the project, e.g., aesthetic or epistemological.
- Consider the type of goals or results the project needs to deliver, e.g., insights into user experience or generalisable results.

Making these positions explicit helps to develop shared standards in BCD and contributes towards the development of a research culture specific to the field with its own epistemological standards. Hence, the emergence of a dialogue around the standards that research in BCD should adhere to for each specific research proposal can be a step forward to the development of a distinct framework for doing research in this context, grounded in a synthetisation instead of a juxtaposition between behaviour science and design.

References

- Andersen, H. (2016). Collaboration, interdisciplinarity, and the epistemology of contemporary science. *Studies in History and Philosophy of Science Part A, 56*, 1–10.

- Ball, J. (2019). The Double Diamond: A universally accepted depiction of the design process. *Design Council, 1.*

- Cash, P. J., Hartlev, C. G., & Durazo, C. B. (2017). Behavioural design: A process for integrating behaviour change and design. *Design Studies, 48*, 96–128.

- Cash, P., Gamundi, X. V., Echstrøm, I., & Daalhuizen, J. (2022). Method use in behavioural design: What, how, and why? *International Journal of Design, 16*(1).

- Choi, B. C., & Pak, A. W. (2006). Multidisciplinarity, interdisciplinarity, and transdisciplinarity in health research, services, education, and policy: 1. Definitions, objectives, and evidence of effectiveness. *Clinical & Investigative Medicine, 29*(6).

- Churchman, C. W. (1967). Guest editorial: Wicked problems. *Management Science, 14*(4), B141–B142.

- Colusso, L., Jones, R., Munson, S. A., & Hsieh, G. (2019, May). A translational science model for HCI. In *Proceedings of the 2019 CHI Conference on Human Factors in Computing Systems* (pp. 1–13).

- Coyne, R. (2005). Wicked problems revisited. *Design Studies, 26*(1), 5–17.

- De Boer, B. (2021). *How scientific instruments speak: Postphenomenology and technological mediations in neuroscientific practice.* Rowman & Littlefield.

- De Boer, B., Te Molder, H., & Verbeek, P. P. (2020). Constituting 'visual attention': On the mediating role of brain stimulation and brain imaging technologies in neuroscientific practice. *Science as Culture, 29*(4), 503–523.

- De Vries, R. A., Haarman, J. A., Harmsen, E. C., Heylen, D. K., & Hermens, H. J. (2020, October). The sensory interactive table: Exploring the social space of eating. In *Proceedings of the 2020 International Conference on Multimodal Interaction* (pp. 689–693).

• Dalsgaard, P. (2017). Instruments of inquiry: Understanding the nature and role of design tools. *International Journal of Design, 11*(1), 21–33.

• Farrell, R., & Hooker, C. (2013). Design, science, and wicked problems. *Design Studies, 34*(6), 681–705.

• Fedlmeier, A., Bruijnes, M., Bos-de Vos, M., Lemke, M., & Kraal, J. J. (2022). Finding what fits: Explorative self-experimentation for health behaviour change. *Design for Health, 6*(3), 345–366.

• Fischer, A. R., Tobi, H., & Ronteltap, A. (2011). When natural met social: A review of collaboration between the natural and social sciences. *Interdisciplinary Science Reviews, 36*(4), 341–358.

• Gaver, W. (2014). Science and design: The implications of different forms of accountability. In *Ways of knowing in HCI* (pp. 143–165). Springer.

• Groeneveld, B., Dekkers, T., Boon, B., & D'Olivo, P. (2018). Challenges for design researchers in healthcare. *Design for Health, 2*(2), 305–326.

• Hallsworth, M., & Kirkman, E. (2020). The future of behavioral insights demands human-centered design. *Behavioral Scientist.* Retrieved from https://behavioralscientist.org/the-future-of-behavioral-insights-demands-human-centered-design/

• Hariton, E., & Locascio, J. J. (2018). Randomized controlled trials—the gold standard for effectiveness research. *BJOG: An International Journal of Obstetrics and Gynaecology, 125*(13), 1716.

• Hartson, H. R., Andre, T. S., & Williges, R. C. (2001). Criteria for evaluating usability evaluation methods. *International Journal of Human-Computer Interaction, 13*(4), 373–410.

• Huutoniemi, K. (2010). Evaluating interdisciplinary research. In *The Oxford handbook of interdisciplinarity* (pp. 309–320). Oxford University Press.

• Jacobs, J. A., & Frickel, S. (2009). Interdisciplinarity: A critical assessment. *Annual Review of Sociology, 35*(1), 43–65.

• Keestra, M. (2017). Metacognition and reflection by interdisciplinary experts: Insights from cognitive science and philosophy. Issues in Interdisciplinary Studies, 35, 121–169. Khadilkar, P. R., & Cash, P. (2020). Understanding behavioural design: Barriers and enablers. *Journal of Engineering Design, 31*(10), 508–529.

- Latour, B. (1987). *Science in action: How to follow scientists and engineers through society.* Harvard University Press.
- Michie, S., & Johnston, M. (2012). Theories and techniques of behaviour change: Developing a cumulative science of behaviour change. *Health Psychology Review, 6*(1), 1–6.
- Michie, S., Van Stralen, M. M., & West, R. (2011). The behaviour change wheel: A new method for characterising and designing behaviour change interventions. *Implementation Science, 6*, 1–12.
- Ministry of Health, Welfare and Sport. (2019). *National Prevention Agreement.* Netherlands: Ministry of Health, Welfare and Sport. Retrieved from https://www.government.nl/documents/reports/2019/06/30/the-national-prevention-agreement
- Moody, L. (2015). User-centred health design: Reflections on D4D's experiences and challenges. *Journal of Medical Engineering & Technology, 39*(7), 395–403.
- Norman, D. A. (2010). The research-practice gap: The need for translational developers. *Interactions, 17*(4), 9–12.
- Oulasvirta, A., & Hornbæk, K. (2022). Counterfactual thinking: What theories do in design. *International Journal of Human-Computer Interaction, 38*(1), 78–92.
- Reay, S., Craig, C., & Kayes, N. (2019). Unpacking two design for health living lab approaches for more effective interdisciplinary collaboration. *The Design Journal, 22*(sup1), 387–400.
- Rittel, H. W., & Webber, M. M. (1973). Dilemmas in a general theory of planning. *Policy Sciences, 4*(2), 155–169.
- Sauro, J., & Lewis, J. R. (2016). *Quantifying the user experience: Practical statistics for user research.* Morgan Kaufmann.
- Tobi, H., & Kampen, J. K. (2018). Research design: The methodology for interdisciplinary research framework. *Quality & Quantity, 52*, 1209–1225.
- Zimmerman, J., Forlizzi, J., & Evenson, S. (2007, April). Research through design as a method for interaction design research in HCI. In *Proceedings of the SIGCHI Conference on Human Factors in Computing Systems* (pp. 493–502).

213

Catelijne van Middelkoop

12. Harmonizing Worldviews: The Transformative Powers of Artistic Research within Design Education

DOI: 10.1201/9781003609766

Introduction

Artistic practices, described as performative acts, possess a unique capability to influence and reshape our perceptions of the world, often affecting us on a profound, moral level (Borgdorff, 2006). This transformative potential becomes especially significant when addressing divergent worldviews, where individuals are typically aware only of their own assumptions about reality (Kottko, 2004). Artistic research serves as an essential tool in bridging these diverse perspectives, acting as a unifying force that brings different viewpoints closer together. As Christopher Frayling (1993) suggests, it is akin to a prelude in music, wherein the composer experiments with harmonies that will later define the main composition.

Although artistic research has historically been intertwined with (applied) art practice, its formal integration into design education in the Netherlands remains limited (Van Middelkoop, 2025). However, this study offers insights that may also be valuable to other educational institutions on all levels (mbo, hbo, wo) whose legitimacy in educating designers is challenged by political, socio-economic and technological changes.

This chapter advocates for greater recognition of the value of artistic research within design education, using a case study of the creative secondary vocational school SintLucas to demonstrate how design researchers can benefit from the methodologies and insights of artistic research, particularly in educational settings. By doing so, it seeks to establish common ground with other contexts where artistic research plays a crucial role, especially in design research processes that prioritize reflective exploration over fixed outcomes and that take place within as well as beyond the creative industries.

Theoretical (Un)framework

Artistic research has gained increased legitimacy and recognition in recent years due to various cultural, institutional, and disciplinary developments. For example, traditional academic frameworks have broadened to recognize that knowledge can be generated and communicated through practice (Borgdorff, 2012), and postmodern

216

and post-structural critiques of objectivity and universal truths (Lyotard, 1979), have increased openness to alternative methodologies, including embodied, performative, and experiential forms of inquiry (Barrett & Bolt, 2014; Biggs, M. & Karlsson, 2011; Haseman, 2006) that argue for the legitimacy of "knowing-in-action" (Schön, 1983).

The creative process as an active form of inquiry is exemplified in both artistic research and Research through Design (RtD). Applied design research aligns naturally with the RtD framework (Stappers & Giaccardi, 2014), as both prioritize creative exploration and knowledge generation through design practice. RtD gained institutional recognition due to its ties to technological innovation and human-computer interaction, benefiting from its alignment with industry needs. However, it has increasingly been evaluated based on functionality, usability, and interaction; criteria not only tied to empirical improvement but also critiqued for reinforcing market-oriented, neoliberal practices. In contrast, artistic research emphasizes ambiguity, critique, and speculative narratives. These approaches have not only gained value in cultural studies and the humanities but have also underscored the field's relevance in addressing urgent global challenges.

Little 'r'

While the establishment of dedicated research programs, practice-based PhDs, and funding for artistic research projects has led to its modest integration into higher vocational and (applied) university education in the Netherlands, the potential of artistic research remains largely overlooked in secondary vocational education (Van Middelkoop, 2022). Yet, the concept of "research with a little r" (Frayling, 1993), deeply rooted in the traditions of art education and practice, closely aligns with the goals of vocational training, which focuses on developing skilled professional practitioners.

Before RtD and artistic research evolved into their current separated forms, Frayling identified three distinct approaches to research in art and design: (1) "research into art and design," (2) "research through art and design," and (3) "research for art and design." Using Frayling's three approaches as a framework

217

for examining contemporary design education in secondary vocational settings, there is a notable lack of historical "research into (contemporary) art and design" education within this context (Van Middelkoop, 2025). In contrast, "research (conducted) for [the purpose of] art and design [production]" is well represented, often involving visual research or other forms of making that directly shape the final outcome of the design process, frequently cited as the primary rationale for engaging in research (Van Middelkoop et al., 2025). However, when the emphasis shifts to the process itself, with the result being more open-ended, new insights can emerge directly from the act of creation.

Both tangible and iterative, the approach in which "research [is conducted] through art and design" holds significant potential, particularly in contexts where making is considered "second nature," such as at creative secondary vocational school SintLucas in the Netherlands. Here, the artistic experience is an active, creative, and aesthetic process where form and content are inseparably linked. This contrasts with other types of implicit knowledge, which can usually be considered and described separately from how it is acquired (Klein, 2010). During research, artistic experiences can arise at various moments, with differing durations and levels of significance. As Klein (2010) notes, this complexity makes it challenging to categorize projects but also enables a dynamic approach to classification: at what stages and in which phases can research be considered artistic? And furthermore, in what contexts?

While some used to argue that artistic knowledge must be verbalized to be comparable to declarative knowledge about facts, concepts, or "what" something is (Jones, 1980) before it could gain value, others claim that the knowledge and therefore its value are embodied in the (semi-)finished products of the creative process itself (Lesage, 2009). As such, it can only be understood through sensory and emotional perception, from which it cannot be separated; artistic knowledge is embodied knowledge (Klein, 2010).

218

Expanding Notions of Knowledge Production

The concept of embodied knowledge emerged in different forms as early as the 1940s through the works of phenomenologists like Merleau-Ponty who said that "the body is our general medium for having a world", but it gained traction across disciplines by the 1950s and 1960s when Polanyi claimed that through *tacit knowledge* "we know more than we can tell" (1966) and further developed in the 1980s and 1990s through cognitive science and reflective practice in design and education (Schön, 1983). Today, embodied knowledge remains central in discussions of artistic and design research, emphasizing the physical, experiential, and material aspects of learning and knowing, emphasizing the relevance of 'making' as a didactic principle (Van Middelkoop et al., 2024).

Context

Although the overarching perspectives through which the world is understood vary across institutions offering design programs in the Netherlands, the design process that students experience shares common elements across different educational levels. Whether in vocational schools (*vakscholen*) and Regional Education Centers (*ROCs*) in secondary vocational education (mbo), universities of applied sciences and art academies (hbo), or scientific education (wo) at general and technical universities (Van Middelkoop & Pescatore Frisk, 2024), certain foundational steps are present in many design methodologies. So, while the idea that the design process is flexible and adaptable rather than entirely universal is a common viewpoint in design theory, it typically follows a general framework. This includes a research and discovery phase, an ideation phase, prototyping, testing and evaluation, and implementation. Depending on context, discipline, and individual approaches, each design— even 'wicked' problem—requires a tailored approach (Buchanan, 1992) along with 'reflection-in-action' (Schön, 1983) allowing for real-time adjustments throughout the process.

This 'reframing' is exemplified in the set-up of the *Practorate Meaningful Creativity*, the first practice-based research professorship (*practoraat*) at secondary vocational school SintLucas, that conducted studies, practical experiments and theoretical (ground)work from August 2021 until July 2024. The practorate began with a profound and open-ended question regarding the future of the organisation it was part of: 'Will our school still [have a right to] exist in 2050?' and the following intention to 'un-blackbox' the overlooked potential of creative Secondary (TYP-O!) vocational education to show its relevance for the future (Van Middelkoop, 2022).

Through iterative adjustments driven by new research insights and unexpected challenges and outcomes, this case study highlights the open-textured nature of design (Van Middelkoop, 2025) and underscores the crucial role of artistic research within an applied design research process.

Methodology

The primary research approach within the practorate was Research through Design (RtD), chosen for its interdisciplinary nature and emphasis on iterative prototyping, which aligned well with the hands-on learning environment of a creative secondary vocational school like SintLucas. However, the concept proved difficult for all stakeholders to fully grasp. While they used the same terminology, terms like "research" and "design" often carried different meanings for different individuals.

Instead of getting bogged down in debates over terminology, we shifted our focus to making and experimented with different ways to engage our stakeholders. Recognizing that not all actions could be meticulously planned in advance, we prioritized thorough documentation of the research process for future reflection and critical analysis; capturing trials, errors, and insights using whichever format or methodology best suited the context.

220

Before we could apply a similar approach to addressing the main research questions that formed the foundation of the three-year research program, we first had to overcome a significant obstacle: obtaining approval for the *practoraatsplan*, which had to adhere to a fixed, predetermined format.

Form Follows Template?

This requirement stems from the fact that in vocational education (mbo), the development of practorates is not yet as extensive or systematically structured as that of lectorates in higher vocational education (hbo) (Andriessen et al., 2023). To enhance the quality of practorates, *Stichting Practoraten.nl* published a white paper in 2020 outlining quality criteria for practorates (Stichting Practoraten.nl, 2020). A national quality committee oversees these standards and provides a framework to ensure consistency and improvement. The committee reviews the plans of emerging practorates (*practoraatsplan*) using an appreciative, development-focused approach to ensure they meet the established criteria.

To have the research design of the *Practorate Meaningful Creativity* reviewed and approved, we were required to submit a standardized *practoraatsplan* template. This form (Practoraten.nl, 2017) was divided into fixed sections following a linear sequence:

- **Principles of the practorate** (e.g., "Briefly and concretely describe the reason for establishing a practorate," and "Describe the goals, outcomes, and impacts using Specific, Measurable, Acceptable, Realistic, and Time-bound criteria.")
- **Phasing and planning** (e.g., "Indicate the phases of the practorate, the planned activities, and the expected results.")
- **Partnership and organization**
- **Monitoring and research**

While such a structure may simplify the review process by offering clear evaluation criteria, presenting it as a strict guide can, as demonstrated in this case, constrain experimentation and innovation by imposing a rigid framework. Instead, we adopted an iterative design process, moving fluidly between problem framing and experimentation. Therefore, the final template was only completed after the research design was firmly established through (reflecting on) designing a series of artifacts.

Speculative Futures Roadmap

An open question about the future can be approached in many ways, depending on the disciplinary focus, agenda, origin of the question, and the availability of data or evidence. To explore this, we employed a *speculative futures roadmap* (Van Middelkoop, 2022) gathering a small but diverse group involved in the development of a Center of Expertise at SintLucas, where the practorate would act as a catalyst. Together, we reconstructed some key events leading up to the decision to establish the practorate, blending a chronological sequence of both recontextualized and previously decontextualized events, current trends, forecasts, unknowns, and uncertainties from professional, personal and even posthumanist perspectives. This process led to the formulation of various rough scenarios and potential outcomes for the practorate in both the short and long term, raising new questions and highlighting potential risks as we projected further into the future. What stood out where the lack of a shared historical awareness, the specificity of some entries compared to others, the differences in proximities of the futures that were projected, the apparent personal 'signatures' of the participants highlighting a wealth of different expertise and interests, and the need for a clear artistic vision, or at least a common goal to strive for.

Position Paper

This vision was captured in a position paper that articulated a clear stance on the future role of creative professionals: "Creativity is essential. By combining our efforts, the collective strength of the creative industry can address the complex challenges of our time, develop alternative future scenarios, and generate societal impact like no other domain. Vocational education (mbo), where ideals are put into practice, lies at the heart of this 'creative consortium.'" In addition to clearly stating the importance of creative secondary vocational education, the position paper aimed to support this perspective with evidence and persuade its audience to understand, accept, and engage with the practorate's vision. Offering a context for a series of provocative memes, it also served as a stepping stone for future speculations.

Wiki

With a clearly formulated mission in place, one immediate obstacle still had to be addressed: the term 'meaningful creativity,' a construct coined from a survey exploring the general concept of creativity at SintLucas. To clarify the position of the practorate bearing this name, a wiki (in Dutch) was created to provide a working definition for the professorship, with the flexibility for future adjustments: "Creativity is meaningful when it can be sustainably applied in a broader societal context" (Practoraat Betekenisvolle Creativiteit, 2021). When the opportunity presented itself to change the name of the professorship, we deliberately decided to stick with 'meaningful creativity' and build on, and make explicit what was already there, instead of starting over again (and again).

Tacit Knowledge

Tacit knowledge, the implicit, intuitive understanding that artists and (most) designers rely on during their work (Schön, 1983), played a significant role in the process of finalizing the *practoraatsplan* and the years that followed after the plan and research objectives were approved. It enabled the practor to improvise and adapt her actions based on the situation, even when the reasoning behind her decisions wasn't always clearly articulated. This ability to improvise proved crucial in navigating the hidden complexities, unknowns, and uncertainties within the organization.

Retrospective Traceability

It also points out the need for the 'retrospective traceability' and 'reflective methodology' (Haarmann, 2024), a "thoughtful methodology" (Haarmann, 2019), which allows the path of knowledge to become clear in hindsight, only after the artistic

223

process has been carried out. This framework aligns with the idea that artistic research generates knowledge in ways that differ from traditional scientific methods, highlighting the unique cognitive potential of art as a form of inquiry and exploration.

Visual Markers

To make the path of knowledge clearer in hindsight, visual markers; tangible, experiential, or symbolic elements that document, inform, and guide the creative and reflective process, play a pivotal role in artistic research. Functioning similarly to boundary objects (Star, 2010), they facilitate communication and collaboration across disciplines, stakeholder groups, and research participants.

By visually documenting iterative changes and prototypes in chronological order, an overview of visual markers serves as a reflective tool for assessing research directions while also communicating findings to stakeholders, including research members of the practorate. This approach enhances reflexivity, fosters co-creation, and encourages participants to externalize their thought processes, improving both internal and external communication.

This "design pedagogy" extends beyond students to all participants in the research process. Moreover, it has the potential to enrich learning by accommodating diverse perspectives and supporting the communication of non-linear research processes. In this way, visual markers are not merely documentation tools but active agents that shape the research journey and its outcomes.

Figure 12.1. Visual markers used in the (Artistic) Research through Design process of the Practorate Meaningful Creativity served as signposts for creative inquiry, aiding in the externalization of ideas, reflections, and discoveries. They traced the development of concepts, tested new directions, and facilitated the communication of findings within a broader research framework. Together, these markers formed a visual narrative, capturing the artistic journey that unfolded over the course of four years.

Analysis

The research project of the Practorate *Meaningful Creativity* was divided into three distinct phases: a 9-month preparatory phase dedicated to establishing the research design, a 3-year research period, and a 4-month reflection and wrap-up phase. Analyzing the visual markers produced throughout the practorate's activities reveals how *Meaningful Creativity* not only navigated the internal research landscape within the educational institution but also enhanced its external positioning.

From the visual data collected between summer 2020 and fall 2024, three key concepts emerge: **the sphere, the hammer, and the black box.** These concepts encapsulate the thematic and methodological insights gained during the practorate's timeline and serve as symbolic representations of its evolving research trajectory and impact.

The Sphere

The discourse of the sphere begins with an 'empty container,' symbolizing both the absence of relevant content at this stage of the process and the vast range of potential directions to explore. As the constructed term meaningful creativity takes shape and becomes a dynamic node within a much larger universe, its interconnections and interdependencies with other stakeholders continue to evolve and expand.

The Hammer

Amid today's rapid technological advancements, we often overlook the simple tools readily at hand. As a symbol of the default, the hammer serves as a powerful reminder of the creative agency that defines us as human; distinct from both other beings and machines. Regardless of how much the world changes, our bodies, through which we engage with and operate these tools, remain our primary medium for experiencing and shaping the world itself.

The Black Box

The black box embodies both a challenge and a source of inspiration. When approached not as an obstacle but as a catalyst for change, its potential applications become boundless. Like the millions of pixels that form an Image on a screen, its meaning depends on whether we choose to decipher the composition or embrace the mystery. While individual elements may feel fragmented or overwhelming in isolation, their true power emerges when they converge, transforming complexity into collective strength.

Findings

Beyond the valuable insights that this case provides into the challenges and strategies of establishing a research professorship in creative secondary vocational education, artistic research can play a transformative role in applied design research by introducing critical, speculative, and reflective practices that challenge assumptions, foster creativity, and expand the scope of design outcomes.

226

Method	RtD Use	Added Value of Artistic Research
Prototyping	Iterative exploration	Critical prototypes and provocations
Speculative Design	"What if" scenarios	Critical fiction, alternative futures
Critical Reflection	Reflection-in-action	Reflexivity, positionality, ethics
Aesthetic Inquiry	Functional aesthetics	Emotional, sensory, symbolic richness
Sensory Engagement	UX and usability	Embodied, multi-sensory design
Participatory Design	User-centered design	Co-creation, critical engagement
Role-Play and Embodiment	Persona design	Movement, bodily experience

Table 12.1. Examples of how Artistic Research increases the Value of Research through Design (RtD).

Further Research

(Re)integrating artistic components into Research through Design (RtD) approaches makes them more critical, reflective, inclusive, and broadly applicable. By shifting the focus beyond problem-solving and short-term objectives, this renewed balance—grounded in the principles of "Research in Art and Design" (Frayling, 1993)—enables the development of aesthetic, social, and ethical interventions that provoke deeper reflection and foster a more profound understanding of the human experience.

A key strength of this approach lies in its emphasis on tacit knowledge, aesthetic inquiry, and speculative exploration. However, integrating artistic research also presents challenges related to rigor, reproducibility, and the dissemination of knowledge. For instance, for visual markers to function effectively as boundary objects, researchers must remain mindful of their interpretive limitations and ensure their use is inclusive and accessible to diverse audiences.

Such limitations highlight the need for further research and methodological innovation, particularly in establishing clearer standards for validating and sharing tacit, embodied, and experiential knowledge within and beyond artistic research contexts.

Conclusion

Artistic research can serve as a critical, speculative, and reflective counterpart of applied design research, expanding the scope of inquiry and deepening the exploration of meaning, aesthetics, and ethics. By integrating artistic research into applied design processes, researchers and practitioners can move beyond utilitarian concerns to engage with larger philosophical, cultural, and speculative questions. This can lead to more thoughtful, innovative, and impactful design outcomes that do not merely respond to immediate needs but also provoke reflection, challenge norms, and imagine new ways of being in the world.

Key Takeaways

To enable artistic research to reach its full potential, several key conditions must be met:

- **(Critical) Making Matters Most:**
  Ideas must be materialized, whether in tactile or digital forms, to be experienced and to create meaningful impact. This necessitates a focus on *critical makership*, where the act of creating is both a process of inquiry and a means of communication.
- **Space for Experimentation and Reflection (Academic x Artistic Freedom):**
  Artistic research requires room for experimentation, including the freedom to reflect critically on both successes and failures. Trust among stakeholders is essential to foster open collaboration. An expert with practical experience in both artistic research and applied design research should lead the process to provide clear guidance and maintain coherence.

- **Shared Quality Standards for Artistic Output and Knowledge Creation:**
  Clear quality standards for artistic outcomes must be established, understood, and embraced by all stakeholders. To uphold these standards, a clear mandate must define who is and remains responsible for their development and oversight.

By meeting these conditions, applied design researchers can not only thrive but also contribute more effectively to innovation, critical inquiry, and creative exploration.

References

- Andriessen, D., Klaeijsen, A., & Baay, P. (2023). *Kwaliteitsontwikkeling van Practoraten: Voorstel voor een waarderende reviewmethodiek.* Retrieved September 16, 2024, from https://onderwijs124. nl/wp-content/uploads/2023/12/Kwaliteitsontwikkeling-van-Practoraten-voorstel-voor-een-waarderende-reviewmethodiek_ Andriessen-Klaeijsen-Baay_2023.pdf
- Barrett, E., & Bolt, B. (Eds.). (2014). *Practice as research: Approaches to creative arts enquiry.* Bloomsbury Publishing.
- Biggs, M. A., & Karlsson, H. (Eds.). (2011). *The Routledge companion to research in the arts.* Routledge.
- Borgdorff, H. (2006). *The debate on research in the arts* (Vol. 2). Kunsthøgskolen i Bergen.
- Borgdorff, H. (2012). *The conflict of the faculties: Perspectives on artistic research and academia.* Leiden University Press.
- Frayling, C. (1993). Research in art and design. *Royal College of Art Research Papers, 1,* 1–5.
- Gross, M., & McGoey, L. (Eds.). (2015). *Routledge international handbook of ignorance studies.* Routledge.
- Haarmann, A., & Borgdorff, H. (2024, April 9). On artistic research methods and the production of knowledge [Panel discussion]. *What Methods Do. Exploring the Transformative Potential of Artistic Research*, Textile Museum, Tilburg, The Netherlands. Retrieved September 10, 2024, from https://www.researchcatalogue.net/ view/2599481/2599482

- Haarmann, A. (2024). The reflective methodology of artistic spatial research. *Handbook of Qualitative and Visual Methods in Spatial Research, 7*, 55.

- Haarmann, A. (2019). Artistic research: Eine epistemologische Ästhetik. transcript Verlag. https://doi.org/10.1515/9783839446362

- Haseman, B. (2006). A manifesto for performative research. *Media International Australia, 118*(1), 98–106.

- Jones, T. (1980). A discussion paper on research in the visual fine arts. *Leonardo, 13*(2), 89–93.

- Kottko-Rivera, M. (2004). The psychology of worldviews. *Review of General Psychology, 8*(1), 3–58.

- Klein, J. (2010). What is artistic research? *Journal for Artistic Research.*

- Lesage, D. (2009). Who's afraid of artistic research? On measuring artistic research output. *ART&RESEARCH: A Journal of Ideas, Contexts and Methods, 2*(2).

- Lyotard, J. F. (1979/1984). *The postmodern condition: A report on knowledge.* Manchester University Press.

- Merleau-Ponty, M. (1965). Phenomenology of perception (C. Smith, Trans.). Routledge & Kegan Paul.

- Polanyi, M. (2009). The tacit dimension. In *Knowledge in organisations* (pp. 135–146). Routledge.

- Schön, D. A. (1983). The reflective practitioner: How professionals think in action. Basic Books.

- Stappers, P. J., & Giaccardi, E. (2017). Research through design. In M. Soegaard & R. Friis-Dam (Eds.), The Encyclopedia of Human-Computer Interaction (2nd ed., pp. 1–94). The Interaction Design Foundation.

- Star, S. L. (2010). This is not a boundary object: Reflections on the origin of a concept. *Science, Technology, & Human Values, 35*(5), 601–617.

- Stichting Practoraten. (2020). *Stappenplan.* Retrieved January 13, 2025, from https://practoraten.nl/stappenplan/

- Van Middelkoop, C. (2025). (Re)making design history: A plea for (artistic) research through design education [Doctoral dissertation]. University of Groningen.

- Van Middelkoop, C., Van Harn, R., Roebroek, D., Van Laanen, K., Koenen, L., Jonkers, S., Van Beers, M., Van de Vijver, N., & Jeuken, M. (2025). *De laatste makers: Een interdisciplinaire zoektocht naar een mogelijke toekomst van creatieve vakschool SintLucas.*

- Van Middelkoop, C., Van Harn, R., Roebroek, D., & Koenen, L. (2024). Maken als didactisch principe (in het mbo) [Workshop]. *Expertisecentrum SintLucas/Practoraat Betekenisvolle Creativiteit.*

- Middelkoop, C. van, & Pescatore Frisk, R. (2024). Concerning apples & oysters. In Joore, P., Overdiek, A., Smeenk, W., & Van Turnhout, K. (Eds.), *Applied design research in living labs and other experimental learning and innovation environments* (pp. 264–290). CRC Press.

- Van Middelkoop, C. (2022). *[Un]Blackboxing the creative potential of Dutch secondary vocational education.Collaboration for Impact.* NADR.

- Van Middelkoop, C. (2022, June). *Speculative futures roadmap. Methods in (the) Making. Research Station Paper, WdKA.*

Nick Verouden, Sophie Vermaning
and Tamara Witschge

13. Navigating Inclusion and Ethical Challenges in Co-creation

DOI: 10.1201/9781003609766

Co-creation and Social Justice

In recent years, European cities have increasingly embraced participation as a key strategy for fostering innovation in the development of socio-technical solutions. These participatory approaches are intended to promote inclusivity by actively involving citizens in shaping technologies that address pressing urban social challenges. Co-creation, as the latest evolution of this approach, advocates for a collaborative process in which stakeholders—such as end users, community members, policymakers, and businesses—are not just consulted but become equal partners throughout the entire product development cycle. In contrast to traditional top-down methods, co-creation emphasizes dialogue, iterative development, and real-time feedback, aiming to ensure that outcomes are not only functional but also meaningful and equitable. In the context of urban digital transformation, co-creation can serve as a critical mechanism for fostering multidirectional exchanges and challenging traditional hierarchies among local governments, businesses, universities, and citizens (Leino, 2020). By redistributing power and prioritizing inclusivity, it becomes a key tool for developing citizen-driven solutions that address diverse needs.

Despite its promise, co-creation in smart city development is often confined to an expert-driven process that reinforces existing power imbalances instead of democratizing technology (Voorberg et al., 2014). Scholars in social justice, participatory design, and pedagogy advocate for a more radical co-creation approach that directly challenges these power structures (Costanza-Chock, 2020; Freire, 2000). They argue that democratizing technology requires not just public involvement, but the sustained inclusion of vulnerable individuals. Co-creation should be transformative—emphasizing bottom-up engagement, addressing systemic inequities, and dismantling hierarchies to achieve lasting social change. Costanza-Chock's Design Justice (2020) exemplifies this, highlighting participatory processes that empower historically excluded communities. By redistributing power and centering marginalized voices, design justice ensures that technology meets specific community needs without reinforcing inequalities. This approach moves beyond simply gathering input, aiming to co-create equitable and transformative solutions. Ultimately, the challenge is not just to make co-creation more inclusive, but to shift power dynamics, placing historically excluded groups at the center to ensure that technology serves their needs and does not perpetuate systemic inequities.

However, while co-creation holds significant democratic and emancipatory potential (Herberg, 2022), particularly in addressing social justice issues, its practical implementation and execution is fraught with challenges. Diverse stakeholder needs, interests, and power dynamics often result in misunderstandings, conflicts, and imbalances, particularly when engaging excluded groups (Jagtap, 2022). These obstacles can undermine both participation and outcomes. While extensive research exists on co-creation, much of it relies on secondary observations, creating a gap in understanding its real-world application (Leino, 2020; Torfing, Sørensen & Røiseland, 2016). This chapter seeks to bridge that gap by exploring co-creation's role in developing citizen-driven socio-technological solutions within smart cities. Focusing on the ethical and inclusivity challenges inherent in these processes, we draw on ethnographic data to illuminate the tensions that may arise in co-creation practices and processes. By exploring these tensions, we demonstrate their significance for design researchers and practitioners, offering insights into how they can better navigate the complexities of co-creation (Herberg, 2022).

CommuniCity: Innovative Solutions Responding to the Needs of Cities and Communities

Our exploration is based on an ethnographic study conducted within the CommuniCity project, a Horizon Europe initiative involving 12 EU partners. The project aims to launch 100 pilots through three rounds of open calls, each designed to address the needs of European cities and marginalized communities. By leveraging co-creation, the initiative promotes collaboration among city officials, tech developers, designers, and marginalized communities, fostering the development and adoption of innovative and locally relevant digital solutions.

235

Pilots were selected through an open-call system in which tech companies proposed challenges and applied for seed money grants. The first round, launched in early 2023, funded 13 pilots in Amsterdam, Helsinki, and Porto. The second round, beginning in September 2023, supported 25 additional pilots, with Aarhus, Breda, Prague, and Tallinn serving as replicator cities. A total of 100 pilots are anticipated by the final round (see Figure 13.1 for overview of the piloting process). These pilots address challenges such as AI translation tools for refugees, VR programs to foster intergenerational understanding, tech solutions for accessible public transportation, and apps encouraging young girls to participate in sports.

CommuniCity Open Calls in brief

| FEB 20 23 | Open Call round 1 Amsterdam, Helsinki and Porto | SEP 20 23 | Open Call round 2 Amsterdam, Helsinki and Porto + 4 other cities | SEP 20 24 | Open Call round 3 Amsterdam, Helsinki and Porto + Other European cities |

| 3 Cities | 18 Pilots | 7 Cities | 39 Pilots | ? Cities | 43 Pilots |

| 12 500 € / pilot | | 12 500 € / pilot | | Up to 15 000 € / pilot |

Figure 13.1. Internal project document that gives overview of the piloting rounds

Our analysis examines 15 pilots in the first and second rounds conducted in the Netherlands between May 2023 and June 2024. Using an ethnographic approach, we employed participant observation, interviews, and document analysis to capture co-creation in practice. We conducted 31 interviews and numerous informal conversations with stakeholders, including tech developers, designers, city officials, and community members. Although we engaged with a broad range of stakeholders, this chapter primarily focuses on designers and tech developers due to their central roles in driving co-creation. Drawing on Tim Ingold's

(2013) perspective of making as a process of correspondence, we view designers and tech developers as deeply interconnected with the products, environments, and people they engage with. Their methods, assumptions, and decisions shape both the co-creation process and the resulting technologies.

In the remainder of this chapter, we examine the role of designers and tech developers in co-creation processes with marginalized communities, emphasizing how their decisions, methods, and interactions shape these processes and influence ethical considerations, inclusivity, and power dynamics in the design and implementation of socio-technical solutions. Conducting fieldwork for this study posed unique challenges due to the diversity and geographical dispersion of the CommuniCity pilots, each of which featured distinct objectives, contexts, and stakeholders. To address these complexities, we adopted a flexible and adaptive ethnographic approach, moving between sites to capture a broad range of insights through shorter, varied, and episodic engagements. This method prioritized responsiveness to diverse contexts and the emergent nature of social interactions, rather than sustained immersion in a single context or setting (Kjaersgaard, Halse, Smith, et al., 2020). By engaging with designers, tech developers, community members, and other stakeholders across multiple pilots, we identified cross-pilot patterns while remaining attuned to pivotal moments and the relational dynamics of co-creation as they unfolded in specific local contexts. This approach allowed us to construct a nuanced understanding of the interplay between the local specificity of the pilots and broader project dynamics.

In the following sections, we examine three tensions observed across the pilots: (i) challenges in understanding communities and their needs, (ii) ambiguity surrounding the concept and practices of co-creation, and (iii) time constraints, budget limitations, and competing priorities. These tensions serve as the organizing framework for our analysis, with illustrative examples provided in each section to offer deeper insights into their implications.

Challenges in Understanding Communities and Their Needs

Our observations reveal a critical tension in the co-creation process, where tech developers and designers struggled to fully grasp the complexities and understand the needs of the communities involved in the pilots. These challenges were rooted in several factors, including a bias toward predefined technical solutions, reliance on assumptions, and an over-reliance on intermediaries to translate community needs. The challenges faced in understanding the needs of the communities involved in the pilots resulted in limited and oversimplified understandings of community needs, ultimately neglecting the diverse and complex experiences of those involved.

A first issue that emerged was the prioritization of technology as the primary solution. From the outset, stakeholders assumed that technology would inherently address community needs, which strongly influenced and ultimately steered the direction of the co-creation process. In selecting pilot challenges, tech-based approaches were favoured without thoroughly assessing the communities' actual needs. For example, in a pilot aimed at improving public transport for the hearing impaired, the initial focus was on technological solutions. However, one participant pointed out that non-tech solutions, such as training staff in social skills, would be more effective. He noted, "If you can convince conductors that it's not the passengers' fault, you've already solved 90% of the problems." This revealed that designers defined the problem through a technological framework, despite the fact that the people for whom it was being developed identified the real issue as lying in the interpersonal dynamics between passengers and staff. Predefined tech solutions narrowed the understanding of community needs, favouring assumptions over genuine, bottom-up engagement.

In addition to predetermining the type of solution to be developed, our observations secondly illustrated that developers and designers often relied on assumptions about the communities' needs, rather than consulting the communities directly. The challenge was exacerbated by the lack of geographical and cultural proximity. Several tech companies involved in the pilots were based outside the Netherlands, creating a gap in understanding the local context. Designers told us how physical distance, combined with language and cultural barriers, hindered effective communication regarding the needs of communities. For example, a pilot with a Danish tech company aimed at helping Dutch youths with criminal records find employment faced challenges due to these obstacles. Relying on online tools and interpreters, communication was often misinterpreted, leading to a design that did not align with local needs. This highlighted the risks of relying on preconceived ideas and the importance of overcoming geographical and cultural barriers to gain a more accurate understanding of the community's needs.

Third, we found that tech developers and designers had difficulty viewing the community itself beyond a narrow, predefined lens. They often framed entire communities and their needs through assumptions, rather than seeking to understand them from the perspectives of those directly involved. Project descriptions and communications frequently used terms like 'hard-to-reach,' 'marginalized,' and 'vulnerable' without sufficient reflection or clarification, oversimplifying the communities and overlooking the structural factors behind their marginalization. This lack of critical examination not only shaped the design process but also framed community needs in a way that neglected the complexities of their lived experiences.

Scholars such as Costanza-Chock highlight the challenges of defining 'community' and ensuring accountability, noting that the unreflective use of terms like 'marginalized' and 'vulnerable' can misrepresent needs, reinforce stereotypes, and perpetuate power imbalances. Such labels imply that communities lack agency, framing vulnerability as an inherent deficit rather than because of structural inequality (Munari et al., 2021). In our study, designers often described communities through this 'vulnerability' lens, reinforcing the idea that they lacked the capacity to engage with more complex, often technological issues. For example, one pilot team hesitated to involve community members in technical discussions, believing the topics were 'beyond their comprehension.' More generally, communities were often referred to as 'target audiences' or 'testing groups,' reducing them to homogeneous user categories and overlooking their internal diversity, power dynamics, and varying priorities. In other instances, communities were not seen as complex and varied groups of individuals with different backgrounds, needs, and perspectives. Instead, they were simplified or reduced to the challenges they were facing, such as poverty, lack of access to services, or other issues. As one tech developer put it, 'The target audience represents the problem.'

A third issue revolves around who can accurately represent the community's needs. Simplified notions of communities can create representation challenges, amplifying certain voices while marginalizing others. Designers often turned to civil society organizations, or 'intermediaries,' as representatives of the community. While these intermediaries were respected and well-connected, they did not always reflect the full diversity of the community's views and needs. For example, in one pilot, a youth professional's perspective was used to represent students' views on vaping, with the students themselves being scarcely consulted. While youth professionals offer valuable insights, their perspective did not fully capture or translate the diverse experiences of the students. Similarly, a designer working on promoting sports for young girls noted that the girls lacked a platform to genuinely express their needs. Instead, the team engaged with coaches who interpreted the girls' needs from an adult perspective, overlooking the girls' actual voices. Over-reliance on intermediaries thus risks overshadowing the community members' direct input, leading to misaligned solutions.

Ambiguity Surrounding the Concept and Practices of Co-creation

A second tension lies in the ambiguity surrounding the concept and execution of co-creation. The conceptual ambiguity and variability in definitions and applications of co-creation across different contexts is widely noted (Voorberg et al., 2014). Costanza-Chock (2020) emphasizes the importance of clear definitions in participatory processes, especially for terms like co-creation. Without a shared understanding among stakeholders—community members, researchers, designers, and policymakers—misunderstandings can arise, expectations may not align, and collaborations can break down. Clear definitions help ensure productive, meaningful efforts and contribute to sustainable outcomes.

During our fieldwork, we observed that the meaning of co-creation varied among stakeholders from the project's outset, though these differences in interpretation were not openly discussed. Interviews revealed that while tech developers and designers acknowledged the value of co-creation, they often viewed it more narrowly than the broader concept defined in the CommuniCity models. Instead of seeing it as true collaboration or co-production, designers framed it as consultation or user testing. It was seen as a series of specific activities and methods, such as user testing or workshops, rather than as a continuous and integral process. For example, community input was usually gathered during the ideation phase, with members invited to test prototypes. In one case, the pilot project team developed an app for expats and conducted online interviews to present it to the target audience. While this approach made sense from a product development perspective, it did not align with the project's objective of creating continuous and collaborative co-production with community members.

The emphasis on user testing was further evident in the use of traditional methods like interviews and focus groups, which often reduced community involvement to simple data collection. In several pilots, short input sessions—sometimes lasting only

241

30 minutes—were conducted, with expert designers primarily responsible for analysis and synthesis. This limited active community participation in decision-making, reflected broader challenges where co-creation was often reduced to user testing. This meant that even well-intentioned activities unintentionally contributed to a form of data extractivism.

During the first round of pilots, differing interpretations of co-creation among stakeholders were not immediately evident to those involved. However, in the second phase of the CommuniCity project, these differences became more apparent, especially during internal management meetings. Researchers, including ourselves, highlighted that the co-creation approach used in the pilots was too narrow and lacked reflexivity and needed refinement. Despite this, proposals to change the approach in the second round were only partially implemented. The continued lack of clarity around co-creation led to growing misunderstandings among stakeholders, making it harder to effectively address concerns. While some adjustments were eventually made—such as modifications to online co-creation workshops for designers and tech developers before the pilots began—much of the pilot design had already been finalized. This left little room to adapt the co-creation process as the project progressed. This leads us to the final tension: the disconnect between the ideal of co-creation and the practical constraints imposed by project organization and coordination.

Time Constraints, Budget Limitations, and Competing Priorities

A third tension arose from the recognition that the rigid pilot framework often limited the co-creation process. Designers and tech developers expressed frustration with the disconnect between the project's specific goals and the broader, more flexible objectives of co-creation, an issue also highlighted in the literature

(Jagtap, 2022). In our study, the limited time and budget allocated for the pilots, combined with the immediate priorities of designers and tech developers, made it difficult for the co-creative process to unfold gradually and organically.

Conversations with stakeholders highlighted how tight deadlines and limited resources severely impacted the quality of co-creation. During the six-month piloting phase, developers and designers faced pressure, particularly when organizing sessions and building rapport with community members. A developer working on an app for citizens with disabilities explained that the rigid schedule left little room for flexibility, leading to delays and missed opportunities for collecting essential community data. As he put it, "The biggest issue we faced was the timing of the pilot. The accessibility data had to be collected in April to complete results and workshops by the end of May." To meet these demanding deadlines, pilot teams often turned to civil society organizations to represent community needs, rather than engaging directly with community members. This shift resulted in prioritizing user testing over authentic community involvement, which further diminished the potential for true co-creation.

Budget constraints created even more challenges, with several developers and designers acknowledging that the limited funds available were insufficient to support meaningful co-creation. As a result, teams sought additional funding, which often diverted attention from the pilot, undermining both the quality of community engagement and the overall development of the tools. Beyond these time and budgetary pressures, aligning the short timelines of the pilots with the long-term strategic objectives of the tech companies added another layer of complexity. For many design companies, the pilots were only one phase within larger, ongoing projects aimed at advancing product development in Amsterdam and other European markets. This created a tension between the project's overarching goals, the business imperatives of the companies involved, and the needs of the communities. As one pilot team founder remarked, "On one hand, we're driven by societal impact, but ultimately, we need to generate revenue." This tension often led companies to prioritize technological

243

development over building strong community relationships, hindering meaningful engagement and collaboration. The misalignment of these goals further undermined the co-creation process, as companies focused on business targets rather than fostering lasting ties with community members. This misalignment meant that while the pilots offered valuable learning opportunities, creating a space for innovation and testing innovative approaches, they did not lead to immediate, tangible results, which postponed the implementation of effective tools for the communities they aimed to support.

Conclusion and Guidelines for Co-creation

In this chapter, we explored the challenges of executing and implementing co-creation in the digital transformation of cities, with a focus on the needs of marginalized and hard-to-reach groups. Drawing on an ethnographic study of the first round of CommuniCity pilots, which aimed to co-create citizen-driven socio-technical solutions, we highlighted the persistent gap between inclusive and ethical aspirations and the tensions that hinder their realization. Our study identified three key tensions in achieving meaningful co-creation: challenges in understanding communities and their needs, ambiguity surrounding the concept and practices of co-creation, and constraints related to time, budget, and competing priorities. These challenges reinforced patterns of exclusion and limited the democratic and transformational potential of co-creation (Costanza-Chock, 2020). To address these issues and promote more inclusive, adaptive, and effective co-creation practices, we propose four guidelines: (i) rethinking co-creation methods, (ii) integrating reflexivity into the process, (iii) creating spaces for change, and (iv) balancing scalability with local relevance. We briefly introduce each of these guidelines here.

Rethinking Co-creation Methods

The first guideline emphasizes that meaningful co-creation requires a fundamental rethinking of methods to ensure they are rooted in the cultural and social realities of the communities involved (Herberg, 2022). Co-creation must begin by prioritizing cultural contexts, informal relationships, and the fluid nature of lived experiences. This involves developing approaches that are better equipped to address the diverse and changing needs, interests, and concerns of community members and stakeholders, ensuring their ongoing and meaningful participation. Without sustained and continuous engagement, communities risk becoming mere cardboard cutouts—visible but not genuinely involved—resulting in their contributions being underutilized and solutions that fail to meet their actual needs (Jagtap, 2022).

A design ethnographic approach offers a pathway forward by rethinking co-creation as a dynamic, reciprocal process grounded in mutual exchange and shared understanding. This perspective treats communities as active participants, shaping the design process at every stage, and frames co-creation as an inclusive, adaptive, and continuous effort. It emphasizes ongoing, active involvement by prioritizing the changing contexts, personal experiences, informal connections, and evolving needs of individuals and communities (Drazin, 2000).

This requires replacing rigid, pre-defined frameworks with adaptive, iterative processes. Drawing on Ingold's concept of correspondence (2018), co-creation should be seen as a collaborative journey, built on relationships and shared discovery, rather than a linear problem-solving task. It involves engaging participants "in their hopes and dreams" (p. 7), fostering mutual learning, and maintaining flexibility. By grounding co-creation in the lived experiences and aspirations of all stakeholders, it evolves from mere problem-solving to a process of shared discovery and creation, rooted in the realities of everyone involved.

Integrating Reflexivity

The second guideline underscores the importance of reflexivity for meaningful co-creation. A key element of "corresponding" is the ability of both designers and participants to critically reflect on the project-specific knowledge and ways of working they bring, much of which is often tacit or intuitive. Our project demonstrates how this approach is vital in addressing the ambiguity surrounding co-creation, where differing interpretations often coexisted without being openly discussed or reconciled. Müller (2000) emphasizes the significance of recognizing and integrating such informal, tacit, and non-verbalized knowledge within design ethnography. By making this knowledge explicit and subjecting it to critical reflection, we not only enrich co-creation practices but also foster openness to dialogue and collaboration.

In our project, we developed the Co-Creation Re-Boot Card Set for the third round of pilots. By prompting participants to critically assess their assumptions, knowledge, and contributions at different stages, the card set fosters an inclusive dialogue where diverse perspectives are explored and valued. This tool serves as one example of how reflexivity can enhance the adaptability and contextual sensitivity of co-creation processes. Another potential strategy could be appointing outside facilitators to guide and support the process, ensuring that all stakeholders remain actively engaged and helping bridge differing viewpoints. Such strategies help cultivate an environment where contributions are not only acknowledged but critically examined, thus directly informing the development of the project.

Including Spaces for Change

As a third guideline, we suggest incorporating flexibility and space for change into project frameworks. The success of reflexive methodologies, such as design anthropology, depends on creating space to challenge entrenched project structures. Rigid frameworks often suppress organic, bottom-up approaches, reinforcing path dependence—where past decisions limit adaptability and lock processes into inflexible patterns that fail to meet evolving needs (Mahoney & Schensul, 2006). Our study demonstrates how strict deadlines, budget constraints, and business-driven objectives undermined the flexibility required for community-driven, adaptive co-creation. These rigid conditions transformed what could have been a dynamic and collaborative process into transactional exchanges, limiting opportunities for genuine engagement, creativity, and mutual learning (Fry, 2019).

246

In the CommuniCity project, this challenge was complicated by the adaptation of tech-driven frameworks from two previous EU digital city initiatives. These frameworks incorporated marginalized communities as target groups without reevaluating the tools and methods to ensure meaningful engagement and genuine co-creation. This reliance on inherited structures created a rigidity that hindered the project's ability to address the unique needs of these communities. To address this, projects should embrace flexibility and iterative processes that allow for ongoing learning and timely adjustments. Our study, for instance, specifically emphasized the importance of time—both in understanding long-term community practices and in facilitating the development of co-creation pilots. Rushed timelines undermine trust and hindered the depth of engagement. Thus, co-creation processes must be flexible enough to align with the rhythms and evolving needs of the community.

Balancing Scalability with Local Relevance

The final guideline emphasizes the need to balance scalability with local relevance. To fully realize the potential of inclusive projects, scalability must not come at the expense of local context. Scalability in large-scale innovation and research projects often prioritizes efficiency and replicability but can lead to standardized solutions that overlook the unique needs of marginalized communities (Hanna & Park, 2020). In our study, this focus on scale restricted flexibility, making it difficult to integrate local insights and adapt solutions to specific community needs. Although the CommuniCity project was grounded in people-centered scaling (Myerson, 2016), its ambitious goal of implementing 100 pilots across the EU within a strict three-year timeline led to rigid processes, leaving little flexibility to adjust plans as needed. For co-creation to succeed, scalability and local relevance must work in tandem: scalability provides structure and reach, while local relevance ensures solutions remain flexible and connected to the lived experiences of the communities they serve. This requires creating possibilities within projects that allow for scaling while remaining responsive to local needs and realities.

By highlighting these guidelines, we hope we indicate how co-creation is a practice that needs continuous attention and adaption to the situation. As such, there is no fixed strategy that can be employed, but rather asks for flexible frameworks and methods—such as those informed by insights from design anthropology. What is needed is a framework that enables reflexivity and continuous, genuine collaboration with community members.

Key Takeaways

Our study offers some key takeaways for design practitioners involved in complex co-creation processes:

- Clarify the meanings of co-creation early in the project to ensure shared understanding and alignment across all involved.
- Ensure that those directly impacted by the co-creation process are actively involved in shaping the problem definition and setting expectations for the outcomes.
- Ensure that both the co-creation process and its outcomes are inclusive, reflecting the changing cultural contexts, needs, and perspectives of all stakeholders involved.
- Design co-creation processes with flexibility, allowing for adjustments and adaptations to unforeseen challenges and evolving needs throughout the project.

References

- Costanza-Chock, S. (2020). *Design justice: Community-led practices to build the worlds we need.* MIT Press.

- Drazin, A. (2020). *Design anthropology in context: An introduction to design materiality and collaborative thinking.* Routledge.

- Freire, P. (2000). *Pedagogy of the oppressed* (30th anniversary ed.). Bloomsbury Academic.

- Hanna, R., & Park, H. (2020). Scaling social impact: Strategies for achieving broader and deeper impact. *Journal of Social Entrepreneurship, 11*(2), 194–212.

- Herberg, J. A. (2022). The critique of co-creation: Democratic dialogue or displaced politics? In F. A. Kluge (Ed.), *Transdisciplinarity: A research mode for real-world problems* (pp. 24–37). Population Europe Secretariat.
- Ingold, T. (2013). *Making: Anthropology, archaeology, art and architecture.* Routledge.
- Ingold, T. (2018). From science to art and back again: The pendulum of an anthropologist. *Interdisciplinary Science Reviews, 43*(3–4), 213–227.
- Jagtap, S. (2022). Co-design with marginalized people: Designers' perceptions of barriers and enablers. *CoDesign, 18*(3), 279–302.
- Kjaersgaard, M. G., Halse, J., Smith, R. C., Vangkilde, K. T., Binder, T., & Otto, T. (Eds.). (2020). *Design anthropological futures.* Routledge.
- Leino, H., & Puumala, E. (2020). What can co-creation do for the citizens? Applying co-creation for the promotion of participation in cities. *Planning Theory & Practice, 21*(4), 507–525.
- Mahoney, J., & Schensul, D. (2006). *Historical context and path dependence.* Oxford University Press, 454–471.
- Müller, F. (2020). *Design ethnography: Epistemology and methodology.* Springer.
- Munari, S. C., Wilson, A. N., Blow, N. J., Homer, C. S. E., & Ward, J. E. (2021). Rethinking the use of 'vulnerable'. *Health Promotion International, 36*(3), 796–804.
- Myerson, J. (2016). Scaling down: Why designers need to reverse their thinking. *She Ji: The Journal of Design, Economics, and Innovation, 2*(4), 288–299.
- Torfing, J., Sørensen, E., & Røiseland, A. (2019). Transforming the public sector into an arena for co-creation: Barriers, drivers, benefits, and ways forward. *Administration & Society, 51*(5), 795–825.
- Voorberg, W. H., Bekkers, V. J. J. M., & Tummers, L. G. (2024). A systematic review of co-creation and co-production: Embarking on the social innovation journey. *Public Management Review, 26*(7), 1333–1357.

249

Banoyi Zuma, Liliya Terzieva, Margo Rooijackers and Inge Vos

14. Matching and Mismatching Worldviews Entangled in Art – Applied Design Research in "Hip-Hop Dance" towards Impact

DOI: 10.1201/9781003609766

Introduction

Historically rooted in a deficit model, the design for individuals with special needs is evolving through Universal Design, which supports reflexivity awareness by promoting a design process that encourages self-reflection and considers diverse needs, especially of individuals with special needs (Mace, 2024). It promotes accessibility and equity by encouraging inclusive, assumption-free design for products, environments, and services, ensuring usability for all.

Reflexivity theories critique traditional views, urging self-reflection on how societal structures shape perceptions of disability, highlighting how designing for, not with, individuals with special needs perpetuates marginalization and "otherness." This study examines how through art and hip-hop dance reshapes inclusion and challenges reflexive biases.

Stichting het Gehandicapte Kind introduced hip-hop dance classes to promote inclusion among children with and without special needs. Rooted in street culture, hip-hop fosters self-expression, community, and equal participation, creating an inclusive environment that encourages a sense of belonging and mutual understanding through improvisation and freedom of movement.

This study explores how hip-hop dance classes, guided by universal design principles, influence perspectives on (mis)matching worldviews towards inclusion/inclusivity. Using the impact generation framework, it evaluates outcomes like improved social interaction, self-confidence, and physical development, alongside performance measures and strategies ensuring inclusive participation and integrating parental and educator perspectives.

The results reveal hip-hop dance fosters inclusion and transformation, improving self-perception in children with special needs, shifting peer attitudes, and positively influencing parents' views of their children's abilities, highlighting the role of inclusive arts in community-building.

Literature Review

Reflexivity Awareness

Reflexivity awareness, from a theoretical perspective, involves critically examining one's beliefs, assumptions, and actions and understanding how personal and cultural biases influence knowledge and social interactions. Theories of reflexivity highlight the reciprocal relationship between observer and observed, urging researchers' and practitioners' continuous reflection to minimize bias and deepen understanding of knowledge and practice dynamics.

Reflexivity has become a central concept in social sciences and other disciplines, highlighting the significance of self-awareness and critical reflection in research and knowledge production. Table 14.1 (see below) examines the major theories of reflexivity, focusing on key thinkers such as Bourdieu, Giddens, feminist scholars, critical theorists, and ethnomethodologists. Each theoretical approach offers distinct insights into how reflexivity operates in both individual and collective contexts.

No	Theory of reflexivity	Specifics
1.	Bourdieu's Reflexive Sociology	This theory is foundational in the field of sociology. It challenges the notion of objectivity in social research (Bourdieu, 1990) and emphasizes critical reflection on researchers' social positions. Embedded in a broader sociological framework it connects individual practices to social structures and highlights the need to address power dynamics, avoiding reproduction of social hierarchies in research.
2.	Giddens' Structuration Theory	This theory integrates reflexivity into a broader framework that highlights the interplay between human agency and social structures. Individuals actively shape and reconstruct social systems through "reflexive monitoring of action" (Giddens, 1986). Here reflexivity is seen as both individual and collective, as people engage in ongoing self-reflection while simultaneously contributing to the (re)construction of social systems.
3.	Feminist Theories of Reflexivity	Feminist theories of reflexivity stress the need to expose and address power imbalances in research by recognizing biases and limitations in one's perspective. Intersectionality (Crenshaw, 1991) further complicates reflexivity by showing how overlapping identities shape experiences, deepening the complexity of reflexive practices.
4.	Critical Theory and Reflexivity	This theory places reflexivity at the centre of the analyses of power and ideology. It emphasizes reflexivity as essential for developing "critical consciousness", allowing individuals to recognize how their thoughts and actions are shaped by dominant ideologies and social structures (Habermans & McCarthy, 1985). It extends beyond individuals, advocating collective efforts to challenge societal norms and drive emancipatory social change.
5.	Ethnomethodology and Reflexivity	Ethnomethodology views reflexivity as the dynamic process where actions shape and are shaped by their context. Social norms are continuously negotiated through reflexive actions, emphasizing how individuals collaboratively create and sustain social reality (Garfinkel, 1991).

Table 14.1. Generic reflexivity theories overview.

Art as Inclusion Enabler

Art has long been recognized as *"a powerful tool for fostering social inclusion"* (Tepper & Ivey, 2008), enabling marginalized voices, promoting social cohesion, and fostering cross-cultural dialogue and challenging social inequalities. Here, art is looked into as an inclusion enabler by creating spaces for marginalized voices, promoting social cohesion, and fostering dialogue across different cultural, ethnic, and social backgrounds.

Art *provides marginalized communities with a platform* to express their unique experiences and perspectives. It disrupts dominant narratives by offering underrepresented communities a platform for expressing unique experiences and perspectives (Jermyn, 2001)

Inclusion through art emphasizes participatory initiatives, enabling marginalized individuals to engage in artistic practices and contribute to cultural conversations. Art helps communities reclaim their stories and visibility beyond traditional social structures (Hooks, 1995).

Art *plays a significant role in also bringing people together across cultural, linguistic, and social boundaries*, promoting dialogue and mutual understanding. Thompson (2014) found that collaborative art projects create shared experiences, fostering empathy. Community art programs, public installations, and festivals unite diverse groups through common creative processes, building social connections.

Art, especially activist art (Bishop, 2012), *addresses social inequalities* by critiquing power structures and promoting inclusion of marginalized groups through participatory practices focused on race, gender, class, and disability.

Art as an inclusion enabler *operates across multiple dimensions*, from providing platforms for marginalized voices to fostering social cohesion and addressing inequalities. As shown through the diverse theories of thought above, it underscores the importance of participatory, community-based, and activist art practices in

255

promoting inclusive societies. Through shared creative processes and the reimagining of public spaces, art helps to bridge divides, challenge dominant narratives, and empower individuals to participate fully in cultural life.

Hip-hop Dance as Inclusive Art

Hip-hop dance, originating in 1970s Bronx, embodies improvisation, creativity, and street culture, rooted in styles like breaking, popping, and locking. It emphasizes expression, rhythm, and individuality, using dance to tell stories and challenge norms (O'Donnell & Snow, 2018). Beyond technique, hip-hop fosters attitude, self-confidence, community, performed in cyphers or battles, blending competition and collaboration Through its evolution, hip-hop dance has become a global phenomenon, influencing other dance forms, while staying true to its origins of innovation, resilience, and cultural storytelling. As a global phenomenon, influencing other dance forms, it remains rooted in innovation, resilience, and cultural storytelling.

According to Schloss (2009), the key elements of hip-hop dance can be summarized in the following way:

- *Breaking (Breakdancing)*: Foundational style with top rock, down rock, power moves, and freezes, featuring acrobatic, ground-based movements.
- Locking: Sharp movements with pauses or "locks," performed to funk music.
- Popping: Rapid muscle contractions creating a jerking effect, synchronized with beats.
- Freestyle: Emphasizes creativity and improvisation, often showcased in battles or cyphers.
- Musicality: Focuses on interpreting rhythms, beats, and syncopation in the music.

Hip-hop is seen as a transformative tool in promoting social justice and inclusion. Projects such as socially engaged hip-hop dance contests encourage participants to reflect on their societal roles and the inequalities present within their communities, using art as a catalyst for dialogue and change.

Art's inclusivity aligns with hip-hop culture, particularly breaking, which embodies principles like "each one teach one," "do it yourself," and "real recognize real" (Zuma & Rooijackers, 2020), rooted in a philosophy of "peace, unity, love, and fun" (Li & Vexler, 2019).

Social Impact

Social impact, as defined by Nicholls and Daggers (2010) refers to the effects of organizational actions on community well-being, encompassing people, the environment, and the economy. It spans non-profits, corporations, and government agencies, aiming for long-term improvements in health, education, environment, and social equity.

Table 14.2 provides an overview of the key aspects of social impact.

No	Key aspects	Essence
1.	Measuring impact	Organizations use quantitative and qualitative metrics to assess how their actions affect communities.
2.	Corporate Social Responsibility (CSR)	Companies engage in CSR activities to make positive contributions to society while balancing profit-making. CSR includes efforts such as philanthropy, ethical labour practices, and environmental sustainability.
3.	Sustainable Development Goals (SDGs)	Many organizations align their activities with the United Nations' SDGs, which focus on tackling global challenges like poverty, inequality, and climate change.
4.	Social Innovation	This involves creating new strategies, concepts, and organizations that address social needs more effectively than current solutions.

Table 14.2. Key aspects of social impact.

Social impact in practice involves three streams (Kania & Kramer, 2011), namely: *1) Education and Workforce Development; 2) Environmental Sustainability and 3) Equity and Inclusion*. This study focuses on *Equity and Inclusion*, using a social impact generation framework based on the social impact canvas (see Figure 14.1).

257

The Design Research Project (the Case)

Stichting het Gehandicapte Kind aims to build an inclusive society where children with and without disabilities grow, learn, play and practice sports together. It emphasizes that exclusion, not special needs, leads to lifelong challenges, supporting initiatives that promote participation and integration for children with special needs.

The Foundation's programmes span Play, Sports, and Learning, with the inclusive dancing project 'Samen Dansen' under Sports, emphasizing self-determination, attitude change, and sense of belonging. Designed to showcase inclusive dancing's feasibility, Samen Dansen targets parents, dance studios, and sports clubs, demonstrating how minimal effort fosters inclusivity. Partnering with hip-hop-minded dance studios, Urban Dance Studio (Venlo) and Solid Ground Movement (Amsterdam), the project used hip-hop styles (breakdance, popping, locking) suitable for diverse abilities. These studios served as experimental spaces to refine the format, aiming to expand nationwide if successful.

To facilitate collaboration between the Foundation and dance studios, and to introduce the Foundation to hip-hop culture and breaking, a 2-hour workshop was held on April 25, 2024.

A six-week dance series, free of charge for participants, ran from May 11 to June 15, 2024, with recruitment from April 15-30 via social media and promotion by the Foundation's ambassador, a Bboy and contemporary dancer with special needs. Placement notifications were sent May 3, and 29 children enrolled, with and without special needs.

To prepare and support the dance instructors during the six-week programme, three online sessions were initiated (see Table 14.3).

The May 8, 2024, session covered five key topics:

• Introduction: Participant details, including age and special needs, were shared.

- Creating the Space: Emphasized on fostering an environment where children could participate freely and express their own dance style.
- Dance Class: The 'Zone of Proximal Development' (Vygotsky, 2012) was introduced as a framework for tailoring instruction to individual needs.
- Reflection: Highlighted the importance of feedback and iterative improvement, given the pilot's "learning by doing" approach.
- Informal Contacts: Stressed the opportunity for fostering interactions among children, parents, and instructors during lessons.

On May 22, 2024, an online session reviewed progress, focusing on dance instructors' experiences after lesson 2 and comparing them to the Foundation's insights from the other inclusive sports project (athletics and skating). On June 26, 2024, a session evaluated learnings from the six-week programmes.

Research Methodology and Design

The research question that is being explored within this study is: How can hip-hop dance classes, based on universal design principles, positively impact ((mis)matching) worldviews towards inclusion/inclusivity?

The methodology employed to study the case above is Participatory Action Research (PAR) as a qualitative research methodology that involves researchers and participants collaborating to understand social issues and take actions to bring about social change (Cornish, et al., 2023).

Participants

No	Type of participant	Number and expertise (if applicable)
1.	Children with special needs	14 children (48.3%)
2.	Children without special needs	15 children (51.7%)
3.	Parents of the children with and without special needs	1-2 parents/family member per child
4.	Dance studios	Urban Dance Studio in Venlo, Solid Ground Movement in Amsterdam
5.	Hip-hop dance instructors	1-2 per dance studio
6.	Experts from the Foundation	2 to 3 during lessons 1 to 5, acting as observers and reflection generators. 4 during lesson 6
7.	Research team	Applied sciences researchers: 3 Applied sciences student assistants: 3 Practitioner: 3

Table 14.3. Design research participants.

Group Composition

Studio	Time	Group size	Special needs	Gender	Age range
Venlo	2 pm-4 pm	9	3	7 boys/2 girls	6-12 yo
Amsterdam	11 am-1 pm	9	5	4 boys/5 girls	8-15 yo
Amsterdam	2 pm-4 pm	11	6	8 boys/3 girls	6-21 yo

Table 14.4. Overview participants per dance class.

Data Collection

The data was collected by means of reflection journals (designed by the research team and filled in by the experts from the Foundation and the student assistants); minutes of the meetings between instructors and the experts from the Foundation; surveys conducted with the parents of the children with and without special needs; informal talks with the children and parents.

Analysis and Findings

Thematic analysis was employed to analyse data from the project's preparatory phase, including the Foundation's communication materials, conversations with dance studio managers and instructors, parent surveys and communiques, the instructor manual, subscription processes, and interviews with parents and children (with and without special needs).

The first finding is the assumption: *fear of exclusion*. Some parents hesitated to share details about their child's special needs, despite the foundation's focus on supporting such children. This reluctance might stem from fear of programme exclusion, a common experience for parents of children with special needs, revealing that certain assumptions and fears persisted despite thorough preparation.

The data collected during implementation and analysed included silent observers' reflection journals, their field notes, regular conversations with dance instructors and parents, and photos of the dance classes.

The findings regarding the process and logistics as viewed by the Foundation, project team, dance instructors and the silent observers can be summarized as follows:

261

No	Finding	Explanatory notes and quotes
1.	Lack of (full) commitment	*"We struggled with no-shows. A single case in Venlo and several cases in both Amsterdam groups. So much so that in Amsterdam we could have easily fitted all children in one group."*, says a member of the project team and a silent observer. One cannot help but think that the fact that the dance classes were free of charge, lowered the threshold for no-show. At the same time, there is the awareness that in some instances a 6-week commitment might be too much for some parents. What is relevant for a future implementation is to allow for a longer recruitment process in order to prevent 'spur of the moment'-enrolment.
2.	Ratio balance	The initial goal was a 25/75% ratio of children with and without special needs. Venlo began with this ratio but shifted to 33/67% after adding one child, remaining close to the norm. In Amsterdam, the ratios fluctuated from 55/45% to 20/80% in the morning and 0/100% in the afternoon, affecting the perception of the dance lessons: *"Whereas in Venlo, it 'felt' like a regular dance class with a few special need children added, in Amsterdam it 'felt' like a special needs dance class"*, as shared by one of the observers in Amsterdam. Consequently, the group composition dictated the dance lesson approach and output. While in Venlo the children, predominantly consisting of children without special needs, were able to learn a choreography created by the dance instructors, in Amsterdam, predominantly consisting of children with special needs, the dance routine was cocreated with the children; all children contributed personal moves to the hip-hop dance routine based on their abilities.
3.	Parent-instructor role(s) in relation to balancing the behaviour	As a foundation committed to the wellbeing of children with special needs, we advocate that every child should be able to express themselves in their own way. The question during the dance lessons when disruptive behaviour occurred whilst parents were present, was: Who is in charge of intervening; the dance instructors or the parent(s)? The learning is that, before dance classes start, rules of engagement need to be thoroughly discussed with parents, so everyone is aware where the responsibilities of the dance instructors towards the child starts and ends.

4.	Age differences	The intended age range was between 6 and 16 years of age. While in Venlo, the age range was 6 to 12 years of age, in Amsterdam the age range was 6 to 21 years of agebased on the fact that in some cases children with (mental) special needs operate on a different age level than their calendar age would suggest. In hindsight, combining primary school children with secondary school children in one group turned out not to be a logical categorization and is not recommended. However, it should be noted that implementing age limits makes it harder to find a sufficient number of children with disabilities in certain age brackets.
5.	Dance class design	In Venlo, children without special needs learned a set choreography, while in Amsterdam, children with special needs required an approach tailored to their individual abilities. Group composition shaped the dance class methods and final deliverables.
		"Here the realization of the necessity of very well-trained and experienced dance instructors and if need-be the initial presence of a pedagogue who will assist the dance instructor on approaching certain types of behaviour, so that the dance design approach can be completely targeted and tailored to the needs of the group", per fieldnotes of the silent observers.

Table 14.5. Findings: process, setup and logistics.

The thorough analysis of the data collected post-implementation (after the dance classes had come to an end) through the fieldnotes of the silent observers and the project team; through the survey the parents filled in; through the final event conversations held with the children with and without special needs; and through the final report, poster, and video material collated by the Foundation, and the transcripts of the conversations with the dance instructors, verifies and even strengthens the hypothesis of the positive impact of the hip-hop dance practices on inclusion and inclusivity as such.

263

Below are insights from children, parents, and instructors.

No	Parent	Insights
1.	Mother, 11-year-old girl with special needs	"I thought the way the breakdance course was organized was very good. The dance instructors did a great job. They do not assume that the children cannot do it, but simply see how far they can get: what are the children able to do by themselves? To what extent do they come up with an adjustment? And what can we offer them if they do need help? Children with disabilities are too often approached as if they cannot do anything."
2.	Mother, 15-year-old girl with special needs	"What we have all experienced with our daughter over the past 6 weeks...I have seen that my child is happy, and that is not a given. She's growing, she's dancing, she's enjoying herself, that's great to see. Thank you!"
3.	Mother, 15-year-old boy with special needs	"It's special to see. I've been looking for ages for a place where my son can dance. I have finally found a place where everyone is human."
4.	Mother, 15-year-old girl with special needs	"My daughter always sat on the couch or stood in one place during dance lessons. She now makes movements that she did not make before. I can see, that you have helped her come out of her shell."
5.	Mother, 8-year-old boy with special needs	"I really enjoyed seeing him shine like that."
6.	Mother, 10-year-old girl without special needs	"My daughter really enjoyed participating. She enjoyed going there every week. At times she really counted down the days."

Table 14.6. Post-implementation insights from parents.

All dance instructors were highly satisfied with the process, highlighting hip-hop's transformative impact. It was not a mere process of instruction and dance mastering, it was a mutual learning experience involving continuous adaptation to specific needs, group dynamics, and the evolving context of interactions.

As the dance instructors stated:

- "When I look at the culture, at the hip-hop and breaking culture, you just see that inclusivity. It is a culture that does not look at what you cannot do, but at what is possible. The solutions that are available, the challenges. That you become the best version of yourself. Everyone can come in and find their own way."

- "Hip-hop is about enjoyment shared with others, learning from each other. One of the most important hip hop values is Each One, Teach One. We think in solutions. Breakin', poppin' and lockin' are very suitable ways to deal with this, because they complement how everyone moves in their unique way."
- "I didn't know what to expect, but I thought, we'll just go with it. These children are smart enough to set their own limits."
- "I have learned to look at the child even more closely."
- "It was challenging but rewarding."

Parents of children with special needs have reported significant emotional and social benefits from these classes. Seeing their children participate in a mainstream activity boosts their confidence and provides them with a sense of normalcy. The inclusive nature of the dance environment also alleviates parental concerns about their children being isolated or stigmatized. Similarly, parents of neurotypical children value the exposure to diversity and learning to interact inclusively, fostering a sense of community.

Overall, hip-hop dance classes serve as a space for all children to express themselves freely, breaking down barriers and fostering a culture of inclusion both in and outside the classroom.

The dance instructors reflected on their experience in the past six weeks, labelling it "challenging but rewarding" and expressed the intention to follow up this experience by adding inclusive dancing to the portfolio of the dance studios.

Discussion and Conclusion

The case study has shown the art of dancing to be an inclusion enabler on different levels. On the level of the young participants, hip-hop dance classes have proven to be an effective tool for promoting inclusion among children with and without special needs. These classes encourage physical activity, creative expression, and social interaction in a shared environment. For children with special needs, the structured yet expressive nature of hip-hop dance offers a way to develop motor skills, coordination, and body awareness in a supportive, non-judgmental setting, and an opportunity to connect to children without special needs. For children without special needs, the classes foster empathy and awareness by exposing them to peers with different abilities, promoting collaboration and reducing stigma. The study suggests that art, especially hip-hop dancing, can surpass physical, cognitive and social boundaries because of its natural appreciation of community, creativity and freedom of expression. These results endorse Tepper & Ivey's (2008) view on art as a powerful tool for social inclusion and Jermyn's (2001) suggestion that art can disrupt dominant narratives and provide alternative viewpoints.

On the level of important stakeholders, such as parents, dance studios owners and instructors, and staff members of the Foundation, collaborating in this project has brought about in-depth reflexivity on the desirability of bringing together children with and without special needs in co-creative settings. Although, the diverse stakeholders were fully committed to the project, some prior assumptions of matching worldviews between parents and the Foundation's staff members were inaccurate and had to be overcome and reframed. This demonstrates that art can promote dialogue and shared understanding by bridging social and other barriers (Thompson, 2014).

On a meta level the case study ignited the following key insights on social impact in practice, related to equity and inclusion (Kania & Kramer, 2011):

- Art is able to reframe initial understanding/worldview towards inclusion and, in particular, the ability to co-exist despite differences on the surface. Art reframes worldviews toward inclusion, emphasizing coexistence despite surface differences.
- Hip-hop and breakdance go beyond them being classified as a physical activity or specific sports, they enable imaginative practices and reflexivity beyond space and time. Hip-hop and breakdance transcend physical activity, fostering imaginative practices and reflexivity across space and time.
- Initial perception towards special needs can be twisted with the support of art and in specific dance. Dance and art challenge perceptions of special needs, promoting inclusion.

- Inclusion expands perspectives, benefiting both those with and without special needs.
- Social impact continuously balances outcomes, strategies, and performance measures.
- Stakeholder engagement and co-design uncover inclusion principles, breaking down silos.

Overall, the Inclusive Dancing project was deemed successful by the Foundation's decision-makers for its impactful results and rich data. Plans to expand to other Dutch provinces are underway. However, limitations in the pilot phase, including participant imbalances and no-shows in Amsterdam, require follow-up research in the next phase to validate or revise initial findings. The study's outcomes strongly support the Foundation's mission to foster an inclusive society where children with and without special needs grow up together.

Key Takeaways

These key takeaways provide actionable insights for designing universally meaningful experiences:

- Hip-hop dance goes beyond being a form of art or a stream of creative industries, it is an empathic way of expressing and interaction design on its own. Once you employ hip-hop dance, it allows you to contextualize the needs and expectations of everyone involved and serves as an invitation to participate and co-design.
- Bringing diverse groups together by design needs to take into consideration the lens and perspective every actor is stepping in with and an initial alignment (also by design) is a prerequisite to the further success/impact desired.
- Although Universal Design aims for inclusivity, specific cultural, social, or environmental contexts also need to be addressed. Therefore, strive for flexibility in the design to accommodate individual needs without sacrificing the inclusivity of the collective experience.

267

- Facilitating co-creation ensures that the design reflects the various needs and abilities of the participants. It allows participants to have a voice and influence decisions. Co-designing with participants fosters inclusion.
- While matching worldviews (i.e. shared goals) can act as a unifying force, mismatches can emerge due to differing expectations, needs, values, or lived experiences among participants. Value mismatches, as they are opportunities to improve the design.

References

• Bishop, C. (2012). *Artificial hells: Participatory art and the politics of spectatorship.* Verso Books.

• Bourdieu, P. (1990). *The logic of practice.* Stanford University Press.

• Bourdieu, P. (2010). *Distinction: A social critique of the judgement of taste.* Routledge.

• Cornish, F., Breton, N., Moreno-Tabarez, U., Delgado, J., Rua, M., de Graft-Aikins, A., & Hodgetts, D. (2023). Participatory action research. *Nature Reviews Methods Primers, 3*(34).

• Crenshaw, K. (1991). Mapping the margins: Intersectionality, identity politics, and violence against women of color. *Stanford Law Review, 43*, 1241–1299.

• Garfinkel, H. (1991). *Studies in ethnomethodology* (Revised ed.). Polity Press.

• Giddens, A. (1986). *The constitution of society: Outline of the theory of structuration.* University of California Press.

• Habermas, J., & McCarthy, T. (1985). *The theory of communicative action: Volume 1: Reason and the rationalization of society.* Beacon Press.

• Hooks, B. (1995). *Art on my mind.* The New Press.

• Jermyn, H. (2001). *The arts and social inclusion: A review prepared for the Arts Council of England.* Arts Council of England.

• Kania, J., & Kramer, M. (2011). Collective impact. *Stanford Social Innovation Review, 9*(1), 36–41.

- Li, R., & Vexler, Y. (2019). Breaking for gold: Another crossroads in the divergent history of this dance. *International Journal of History Sport, 36*(4–5), 430–448.
- Mace, R. (2024). What is universal design? Retrieved from The Universal Design Project: https://universaldesign.org/definition
- Nicholls, A., & Daggers, J. (2010). *The landscape of social impact investment research: Trends and opportunities.* University of Oxford & Saïd Business School.
- Olmos-Vega, F., Stalmeijer, R., Varpio, L., & Kahlke, R. (2023). A practical guide to reflexivity in qualitative research: AMEE Guide No. 149. *Medical Teacher, 45*(3), 241–251.
- Schloss, J. (2009). *Foundation: B-Boys, B-Girls, and hip-hop culture in New York.* Oxford University Press.
- Social Enterprise Institute. (2023). *Social impact assessment canvas.* Retrieved from https://www.ed.ac.uk/sites/default/files/atoms/files/social_impact_canvas_3.pdf
- Tepper, S., & Ivey, B. (Eds.). (2008). *Engaging art: The next great transformation of America's cultural life.* Routledge.
- Thompson, N. (2014). *Seeing power: Art and activism in the twenty-first century.* Melville House.
- Vygotsky, L. (2012). *Mind in society: Development of higher psychological processes.* Harvard University Press.
- Zuma, B., & Rooijackers, M. (2020). Uncovering the potential of urban culture for creative placemaking. *Journal of Tourism Futures, 6*(3), 233–237.

Kees Greven, Daan Andriessen

15. The Use of a Research Paradigm Workshop for Embracing Different Worldviews in Practice-Based Research

DOI: 10.1201/9781003609766

Introduction

Practice-based research at Dutch universities of applied science is closely linked to practice and involves direct collaboration with professionals. It not only produces knowledge but also promotes professional learning and develops products and systemic change (Franken et al., 2018; Vereniging Hogescholen, 2021). Because of this multiplicity of research goals and the collaboration they require, it is inherently of an interdisciplinary nature, involving various stakeholders with diverse worldviews and disciplinary backgrounds.

Design (research) projects aim to achieve collective goals, such as designing tools, environments, systems, or developing general design principles. These goals require mutual understanding and agreement on what constitutes good research, good design, valid evidence, sufficient results, and ethical behaviour. Disciplinary standards, cultural, and personal beliefs influence how these questions are answered.

To address potential differences among designers, researchers, and other stakeholders, it is essential to pose these in the first place. The literature on boundary crossing highlights that productive collaboration requires designated time to explore and discuss mutual differences (Aagaard-Hansen & Ouma, 2002; Gavens et al., 2018; Hord, 1981). One method to achieve this is through a workshop developed at Utrecht University of Applied Science (HU), which explicates differences in research paradigms and worldviews.

This chapter aims not to provide comprehensive evidence for the workshop's effectiveness but to underpin its reasoning and illustrate its potential. It will first describe the workshop, its rationale, and methodology. A case study will subsequently showcase how the workshop fosters collective reflection and interdisciplinary collaboration within an ongoing co-design project combining design research and healthcare disciplines. Finally, key takeaways for design research professionals will be presented.

272

Introducing the Workshop

At the HU, the research group for epistemic agency (Lectoraat Onderzoekend Vermogen) has developed a workshop that specifically addresses questions surrounding research paradigms. It is used in various contexts, including PhD orientation courses, professional doctorate courses and other research support sessions, as well as in research projects. Although the layout of the workshop may vary from context to context, the content, goals and underlying rationale remain largely the same.

The workshop aims to promote individual and collective awareness of the possibility and legitimacy of different research paradigms. Participants are invited to delve into the nuances and various shades of their different views. Rather than trying to resolve these differences, a mutual understanding is gained that helps in embracing the different worldviews. This can be helpful in (interdisciplinary) collaboration and communication with other professionals or stakeholders. Furthermore, it can deepen the understanding of one's own motivations and choices in practice-based research.

The workshop exemplifies the research group's aim to support practice-based research at the HU that is both methodologically rigorous and practically relevant (Smith et al., 2013). In various iterations, the workshop grew out of different requests for extra grounding and support on the topic of philosophy of science and methodology at the HU. Over time, the workshop has solidified into a tool that has been implemented as a standard part of PhD and PD supervision, and has been used on request in various research projects.

Generally, it is conducted at the start of a project to form a solid foundation for collaborating across research paradigms. The participants are usually involved in research, but other professionals that are part of a project can equally take part in exploring philosophical positions.

The workshop is based on two central tenets: I) Rather than identifying and naming one's research paradigm, the underlying continuums that constitute them are explored to create a common understanding and shared language to talk about different viewpoints. II) Moving together through a physical space creates a deeper sense of connection and mutual respect and understanding.

Tenet I refers to the idea that the constituent characteristics of research paradigms can be conceived of as forming an associated continuum (Lincoln et al., 2011; Moon & Brackman, 2014). For example, ontologies may be ordered from realist to relativist, and epistemologies may be ordered from objectivist to social constructivist to subjectivist, and so on. Rather than figuring out which research paradigm one is ultimately a part of, understanding and sharing the subtleties of these different positions creates a common language to collectively discuss individual beliefs and preferences. Furthermore, it promotes a certain paradigmatic flexibility which is of considerable utility in an interdisciplinary, practice-oriented context.

Tenet II expresses the belief that cognition and collective learning are in part embodied phenomena. According to the French philosopher Merleau-Ponty, our body is an essential tool in making sense of the world. We collectively gain knowledge because our bodies inhabit the same physical space (Clark, 1998; Van Dijk, 2013; Merleau-Ponty, 1945). This idea is further elaborated on by Bart van Rosmalen in his musal perspective (muzisch perspectief) on professional learning (Van Rosmalen, 2016). From this perspective, the body in space and sensory experience play a fundamental role in shaping our beliefs. Actually moving through a physical model and seeing other people taking up different positions consolidates the legitimacy of every different position.

An example of two continua that are often used in the workshop are part of a model that was developed at the Avans University of Applied Sciences (Van der Auweraert & Niessen, 2024). This model subdivides a plane according to two orthogonal axes, pertaining to two spectra within the philosophy of science: the methodological question of whether the research is geared more towards understanding or intervening, and the epistemological question of whether a researcher is looking for objective versus interpretative knowledge, i.e. the extent to which a researcher ought to be distanced from the research object. These two axes then form a playing field, if you will. The central idea that the workshop utilizes, is that each position on the resulting plane constitutes a legitimate vantagepoint, associated with different research paradigms. Taking this playing field as a whole thus promotes personal reflection and mutual respect, as it unites the different viewpoints in one space.

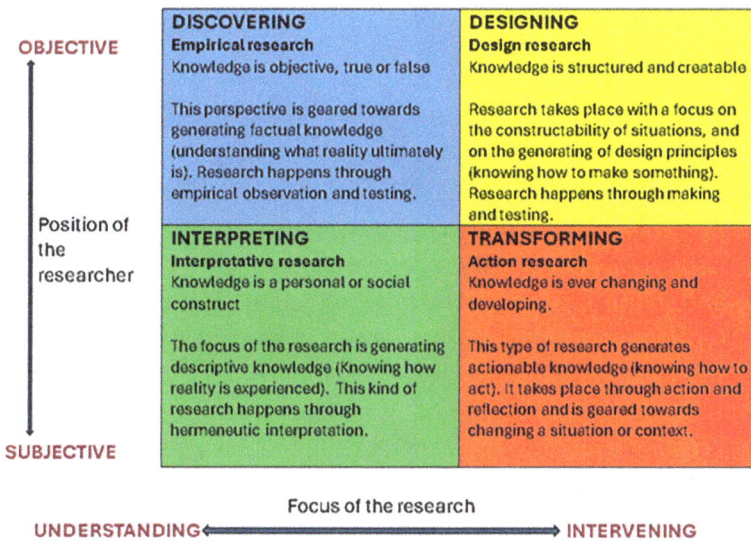

	DISCOVERING	DESIGNING
OBJECTIVE	**Empirical research** Knowledge is objective, true or false This perspective is geared towards generating factual knowledge (understanding what reality ultimately is). Research happens through empirical observation and testing.	**Design research** Knowledge is structured and creatable Research takes place with a focus on the constructability of situations, and on the generating of design principles (knowing how to make something). Research happens through making and testing.
Position of the researcher	**INTERPRETING** **Interpretative research** Knowledge is a personal or social construct The focus of the research is generating descriptive knowledge (Knowing how reality is experienced). This kind of research happens through hermeneutic interpretation.	**TRANSFORMING** **Action research** Knowledge is ever changing and developing. This type of research generates actionable knowledge (knowing how to act). It takes place through action and reflection and is geared towards changing a situation or context.
SUBJECTIVE		

Focus of the research

UNDERSTANDING ⟷ INTERVENING

Figure 15.1. Avans quadrant model.

Although this is a useful way of subdividing a plane for addressing some core philosophy of science issues, they are by no means the only axes that can be defined. As will be showcased below, other spectra relevant for healthcare/design collaboration may, for instance, pertain to structuring, that is, working pre-structured and theory-driven or openly and inductively, or to control, that is, in controlled environments versus experimenting in the field (Zielhuis et al., 2020). Other examples of philosophical spectra are whether one is looking for generalizable facts or context specific knowledge, or whether one strives to be ethically neutral, emancipatory or even activist.

275

Moral position of the researcher

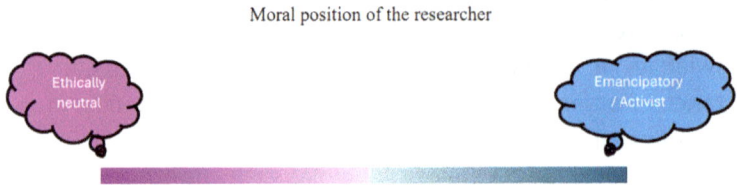

Figure 15.2. Example of philosophical spectrum.

Based on these ideas, the workshop follows the following format:

- Choosing the philosophical gradients that the group would
  like to work on. The Avans model is usually the starting point
  in the context of PD or PhD supervision, to get an overview of
  various research paradigms in one space. In the context of a
  particular project, one may, however, want to focus on specific
  issues relevant to that group or that instance of collaboration,
  and so the axes are defined through discussing points of
  tension, uneasiness or worry together with the participants.
  The facilitator helps to define the terminology and the concrete
  spectra on the basis of what is being shared.
- Creating a physical representation of the relevant concepts
  in space. The quadrants of the Avans model are represented
  by printed descriptions on the floor. The spectra defined with
  the participants may be represented through a line of duct
  tape with descriptors on either side. The crucial point is that
  this representation is not binary, but draws out a continuum
  through which there is space to move. The facilitator has to
  make sure everyone understands the different points within the
  continuums.
- Choosing a position on the plane or on the line. At this step,
  the facilitator invites the participants to actually walk together
  across the physical space and explore the different options.
  Participants are encouraged to speak their initial thoughts and
  intuitions. Not only is this a way of inspiring personal reflection,
  it becomes a group dynamic exercise too. One chooses a
  position relative to their own beliefs and intuitions, and relative
  to the other participants as well.

- Sharing and discussing which personal, professional and/ or academic motivations lead one to occupy this particular position. These motivations may for instance pertain to personal beliefs about ethical behaviour, convictions as to what constitutes professional action, or academic opinions rooted in certain disciplinary backgrounds. Some of these may be quite explicit and others may be held more tacitly. In discussing these considerations, The facilitator encourages everyone to clarify both their implicit and explicit motivations, and makes sure everyone gets heard.
- Invite people to move through the physical representation in order to explore different philosophical standpoints. As the different positions on the line or plane are increasingly developed through the groups dialogue, participants may want to change their position relative to the playing field or relative to other participants. They are thus invited to move through the physical space once more.
- Reflecting on the insights and questions that arose, supported by relevant terminology from the philosophy of science. During the final step of the workshop, participants are invited to reflect together on the differences between their points of view and what these entail for their research and collaboration. The facilitator helps with interpreting what is being said in terms of philosophical debates. In this way, the insights gained from the discussion can be better related to existing literature and academic parlance. Furthermore, participants become more aware of the differences and similarities in the group and are able to express these more articulately.

Case study: A healthcare and design research project at the HU

To bring the workshop to life and to illustrate how it functioned in a specific design research context, let's focus on a particular case in which the workshop was used. The workshop was applied in an ongoing HU co-design project called PEBBLES, that combines healthcare and design research disciplines. The HU research groups on co-design and physiotherapy work together with stakeholders in the field. Before delving into the case study itself, the methodology will be briefly outlined.

277

Methodology

For this case study, the main author of this chapter conducted semi-structured interviews with two of the participants of the PEBBLES project from both HU research groups involved, and analyzed the personal notes and reflections of the participants present at the workshops. As a member check, a PEBBLES researcher and a facilitator provided feedback on this chapter. During the interviews, the two central tenets alluded to above were taken as the defining characteristics of the workshop. Questions thus pertained, amongst other things, to the embodied nature of the exercises, and the way in which philosophical issues are presented as spectra.

For the analysis of the data and the presentation of the case study below, the CAIMER framework was used (Blom & Morén, 2010; Blom & Morén, 2011). This framework distinguishes Context, Actor, Intervention, Mechanism and Results in order to get a handle on how a given intervention is effective. Through explicating the underlying mechanisms, an explanatory chain may be drawn between the concrete activities of the workshop and the observed results (Pawson & Tilley, 2004).

Mechanisms refer to the psychological, cognitive and group dynamic processes that are initiated through the interventions as part of a workshop. These processes are not directly observable as such, but rather constitute underlying explanations as to how the specific elements of the workshop are effective. The outcomes are the effects on the individual participants and on the project as a whole that result from these mechanisms.(Blom & Morén, 2010; Blom & Morén, 2011) In this way, the black box between the concrete interventions and the desired outcomes may be rendered more transparent.

The case study will be presented as follows: first the wider context of the PEBBLES project will be outlined, followed by a description of the intervention, i.e the workshop itself. Subsequently, the findings will describe the role of the facilitator (actor), the principles that made the workshop effective (mechanisms), and the outcomes experienced by the participants (results). The take-away points presented at the end of this chapter will be generalized insights drawn from prying open the black box.

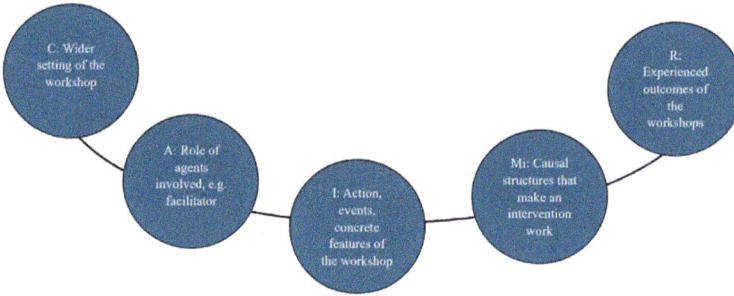

Figure 15.3. CAIMER framework.

Context

The PEBBLES project (Parental Beliefs concerning their Baby, Lifestyle, and Experience Study) is an HU co-design initiative focused on developing tools to map Parental Beliefs & Practices regarding the motor development of children aged 0-2 years. These tools are intended for use in pediatric physiotherapy practice and intervention studies.

Pediatric physiotherapists support infants with motor development concerns, which are crucial for cognitive, language, and physical activity development. Effective interventions require an understanding of Parental Beliefs, which shape Parental Practices. However, physiotherapists often face challenges in assessing these beliefs due to a lack of suitable instruments.

The project aims to create a toolbox to help physiotherapists map parents' thoughts and actions. This toolbox is developed iteratively with input from co-designers, Living Labs of six SME pediatric physiotherapy practices, parents, and researchers. The project emphasizes an interdisciplinary, collaborative approach, focusing on the impact of innovations on both parents and physiotherapists.

279

Additionally, physiotherapists gain experience and skills in design processes. The project highlights the importance of bridging the gap between research and practice, raising methodological and ethical questions.

PEBBLES started in August 2022 and will end in June 2025. The research paradigm workshops were organized at the start of the project, in March and May of 2023, stemming from a collective desire among the HU researchers to explicitly spend time on investigating and promoting the collaboration between the different disciplines, and between the researchers and the other stakeholders, c.q. the pediatric physiotherapists and the parents.

Intervention: The Workshops with PEBBLES

In both workshops, PEBBLES researchers from the HU research groups Co-design and Physiotherapy were present, along with three facilitators from the HU research group Epistemic Agency. The facilitators helped define relevant axes, guided the group through the workshop steps, and provided foundational interpretation.

Based on previous interdisciplinary co-design experiences, PEBBLES researchers anticipated potential issues regarding stakeholder involvement in the research and design process. They suspected differing views on what constitutes good science and collaboration among HU researchers and aimed to address these themes from the get-go. With the facilitators' help, participants formulated the underlying starting question: "How do we analyze our data in such a way that everyone is taken into account?" This question covered both methodological dilemmas and ethical considerations regarding stakeholder expectations.

The workshop continued with participants defining key moments of disagreement or awkwardness related to worldviews or research paradigms. These moments were discussed, clustered, and analyzed with the facilitators, forming the basis for defining several dichotomies. Multiple axes representing potential points of disagreement were defined, including the following:

280

- The extent to which a design research project should be structured in advance or left open to accommodate unforeseen data.
- B) The degree to which professionals and other potential end users may be burdened by involving them in co-creation and testing of new tools.

Axis A addresses how much structuring is needed beforehand in research and design processes, including the structuring of tasks and activities and the extent to which a theoretical lens should be predetermined. These issues reflect the academic and disciplinary backgrounds of the participants, spanning natural and social sciences and design disciplines.

One end of the spectrum views scientific research as hypothesis testing within a theoretical framework, while the other end portrays research as capturing the complexity of the real world through openness and abductive problem-solving (Dorst, 2011).

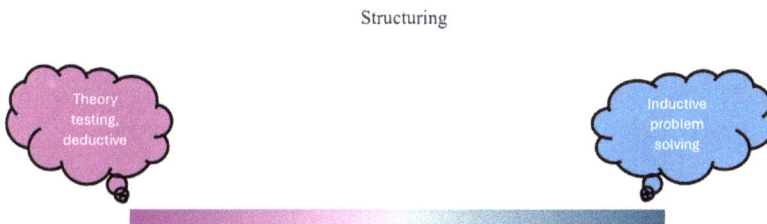

Structuring

Theory testing, deductive

Inductive problem solving

Figure 15.4. Axis A

Axis B represents the ethical dilemma between shared ownership and prudence towards participants. Involving parents and professionals early on promotes shared ownership and valuable insights but may burden stakeholders with testing unfinished prototypes.

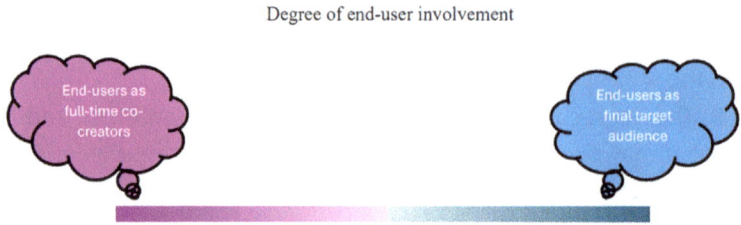

Figure 15.5. Axis B

Findings

Analyzing the PEBBLES case suggests that the design of the workshop according to the two tenets mentioned above helps to bring about a number of results through the effects of a few different mechanisms. In order to ensure the effectiveness of the intended mechanisms, facilitators play an important role in guiding the workshops.

Actor: The Role of the Facilitator

Facilitators play various roles in supporting the execution of the workshop. As mentioned in the description of the workshop's format, facilitators guide the process and make sure the groups dialogue is democratic and fruitful. In the case of PEBBLES, one facilitator specifically safeguarded this process, and thus promoted the mutual respect gained from the embodied experience.

Furthermore, facilitators aid in defining the relevant terminology to work with and help to interpret the different viewpoints. To this end, they relate a number of relevant philosophical concepts to one another and to the points expressed by the participants. This adds extra substance and grounds the learning experience in a wider context. For this, it turned out to be quite valuable that one of the facilitators in the PEBBLES case had a background in philosophy of science.

Mechanisms

The following mechanisms follow from the workshop's tenets:

- Embodied Visualization: The physical representation visualizes the different positions, thereby reifying the abstract dynamics in the group. In other words, it makes the abstract more tangible and able to be experienced.
- Legitimization: The embodied experience happens before any explicit judgement. As proposed in tenet II, this fosters a legitimization in which the different positions can exist side by side, without mutual attempts at convincing the other. By recognizing and acknowledging the differences, it becomes possible to view them as complementary.
- Increasing awareness: Explicating the various viewpoints boosts awareness of both one's own individual position, as well as of the relations within a group. It is made explicit that norms are not universal. Moreover, because differences are, in accordance with tenet I of the workshop, not reduced to black-and-white dichotomies, but represented as many different adjacent shades, one becomes more aware of the subtler differences.
- Team building: It builds team cohesion by focusing together on something else than the project itself. Taking the time together to explicitly address the issue of collaboration shows a mutual willingness to get closer. This is a way to work together in a substantial manner while focusing on something else than the project. Instead, the focus is shifted to an interpersonal level, by occupying the same physical space, and by exploring philosophical questions together.

Outcomes

In reflecting on the sessions, the following outcomes were mentioned as resulting from the workshops. The outcomes pertain to the culture and characteristics of the consortium and to the way it balances differences within the group.

- Language awareness: There is more alertness and awareness in the team of the differences in language. When working across paradigms, one word can carry different meanings, and one and the same phenomenon may be referred to differently by different people. "By recognizing that not everyone means the same by co-creation, prototyping or theoretical framework, miscommunication can be avoided more easily."

283

- Openness: In the team, implicit presuppositions get explicated more easily as a result of the workshop, which reduces confusion later on. By having dedicated time specifically for fostering collaboration, a common culture or group dynamic is formed that normalizes openness, for instance, about what constitutes good research. "You know more precisely where you need one another, and how to utilize everyone's strengths!"
- Mutual respect: A deeper realization that everyone brings their own life experiences, expertise and value to the project. Behind each position on the line hides a wealth of personal experiences, professional knowledge and academic training. The workshop aided in establishing this culture of respect. "You realize that everyone stands on the line as a human being, with unique life events that brought them where they are..."
- Professional learning: The workshop promoted an in depth discussion about the different options of including the professionals and end users in the design process, which lead to a more daring way of experimenting with this strategy. As a result, both the researchers present at the workshop, as well as the physiotherapists gained new learning experiences when it comes to co-designing a product. "In the project, we felt more comfortable truly including the physiotherapists as co-designers."

Conclusion and Discussion

The PEBBLES project presented in this chapter provides an example of an interdisciplinary design project, as part of the practice-based research typical for the universities of applied science. It has to deal with different worldviews because of the collaboration between healthcare and design research disciplines, and because of the close collaboration with professional practice. Holding a research paradigm workshop can help balancing out these different worldviews while not trying to resolve the differences through argumentation. Rather, identifying the deeper continuums that underlie these worldviews through an embodied way of working fosters collaboration whilst doing justice to the complexities of anyone's position. The outcomes that can be seen in the case of the PEBBLES project both pertain to the way of collaborating within the team directly, as well as to the development of the project itself. Through mechanisms such as increasing awareness and teambuilding, the team members became more open towards one another and more accustomed to explicating issues surrounding differences in language and worldview. Furthermore, the legitimization and collective exploration of different viewpoints contributed to the team's willingness to include the stakeholders more directly in the design process. This led both the design researchers and professionals to gain valuable insights about co-designing a product. For example, the design researchers learned that burdening the end user with unfinished products can be done responsibly and practitioners gained experience with designing a product. This led to more shared ownership over the final outcome.

The two central tenets of the workshop in part determine the mechanisms through which it has an effect on the participants. The increase of awareness about the various possible philosophies is enhanced as a result of the way in which dichotomies are presented as gradients, with several subtler differences. Furthermore, the embodied nature of the group exercise builds mutual respect and team cohesion by moving together through a shared space. Moreover, it entails realization that visualizing different worldviews is not just observable from a distance, but rather an object of direct experience.

As mentioned in the introduction, the aim of this chapter has not been to prove the effectiveness of the workshop, but rather to illustrate its potential and illuminate the underlying mechanisms that characterize it. As such, the study is limited to a single case study. In the case of the PEBBLES project, only HU researchers took part in the workshop. Though the workshop had an effect on how they collaborated with the professional stakeholders, those activities were not part of the actual workshop.

In conclusion, the participants of the PEBBLES project who attended the workshops agreed that organizing the workshops contributed to the success of the project. Moreover, the format of the workshop, i.e. the two tenets described above, has been taken

285

up by the project's participants to include it in other meetings as well. On this basis, we therefore recommend interdisciplinary design research projects to include a workshop like this at the start of the program to help with embracing different worldviews.

Key Takeaways

The following points, ordered according to the CAIMeR logic, may be taken away from the above:

- Context: Early on in interdisciplinary design research, use paradigm workshops to clarify and address differences in worldview.
- Actor: Engage a facilitator knowledgeable in the philosophy of science to mediate between diverse worldviews and paradigms.
- Intervention: Develop workshops that emphasize philosophical viewpoints as a spectrum and use shared embodied experiences to build mutual respect and understanding.
- Mechanisms: Utilize embodiment, visualization, legitimization, awareness-raising, and team-building activities to promote the inclusion of diverse worldviews.
- Outcomes: Aim for outcomes such as increased language awareness, openness, and mutual respect, viewing differences as gradients rather than binary oppositions.

References

- Aagaard-Hansen, J., & Ouma, J. H. (2002). Managing interdisciplinary health research: Theoretical and practical aspects. *International Journal of Health Planning and Management, 17*(3), 195–212. https://doi.org/10.1002/hpm.671
- van der Auweraert, A., & Niessen, T. (2024). Mind the gap. *Tijdschrift voor Hoger Onderwijs, 42*, 1–17.
- Blom, B., & Morén, S. (2010). Explaining social work practice—the CAIMeR theory. *Journal of Social Work, 10*(1), 98–119.
- Blom, B., & Morén, S. (2011). Analysis of generative mechanisms. *Journal of Critical Realism, 10*(1), 60–79.
- Clark, A. (1998). Embodiment and the philosophy of mind. *Royal Institute of Philosophy Supplements, 43*, 35–51.

- Dijk, van, J. (2013). Creating traces, sharing insight : explorations in embodied cognition design. [Phd Thesis] Technische Universiteit Eindhoven. https://doi.org/10.6100/IR759609
- Dorst, K. (2011). The core of 'design thinking' and its application. *Design Studies, 32*(6), 521–532.
- Franken, A., Andriessen, D., Van der Zwan, F., Kloosterman, E., & Van Ankeren, M. (2018). *Meer waarde met hbo.* Den Haag: Vereniging Hogescholen.
- Gavens, L., Holmes, J., Bühringer, G., McLeod, J., Neumann, M., Lingford-Hughes, A., Hock, E. S., & Meier, P. S. (2018). Interdisciplinary working in public health research: A proposed good practice checklist. *Journal of Public Health, 40*(1), 175–182.
- Hord, S. M. (1981). *Working together: Cooperation or collaboration.* ERIC.
- Lincoln, Y. S., Lynham, S. A., & Guba, E. G. (2011). Paradigmatic controversies, contradictions, and emerging confluences, revisited. In *The Sage handbook of qualitative research* (4th ed., pp. 97–128). Sage Publications.
- Merleau-Ponty, M. (1945). *Phénoménologie de la perception.* Éditions Gallimard.
- Moon, K., & Blackman, D. (2014). A guide to understanding social science research for natural scientists. *Conservation Biology, 28*(5), 1167–1177.
- Pawson, R., & Tilley, N. (1997). Realistic evaluation. SAGE Publications.
- Van Rosmalen, B. (2016). Muzische professionalisering: Publieke waarden in professioneel handelen (1st ed.). BoekWerkUtrecht.
- Smith, G. J., Schmidt, M. M., Edelen-Smith, P. J., & Cook, B. G. (2013). Pasteur's quadrant as the bridge linking rigor with relevance. *Exceptional Children, 79*(2), 147–161.
- Vereniging Hogescholen. (2021). *Praktijkgericht onderzoek als kennisversneller: Strategische onderzoeksagenda hbo 2022–2025.* Den Haag: Vereniging Hogescholen.
- Zielhuis, M., Van Gessel, C., Van der Lugt, R., Godfroij, B., & Andriessen, D. (2020). Grounding practices: How researchers ground their work in create-health collaborations for designing e-health solutions. In *Proceedings of the 6th International Conference on Design4Health* (Vol. 3, pp. 142–152). Sheffield: Sheffield Hallam University.

287

Part 4: Beyond Solutionism, Design without End

"Embedded in the core of design is an assumption of make ability— the notion that no matter how convoluted the situation, creative possibilities always exist."

Guido Stompff, Tomasz Jaskiewicz,
Troy Nachtigall, Iskander Smit

16. Design in the Real World: Facilitating Collective Learning through Design

DOI: 10.1201/9781003609766

Introduction

Two years ago, in an economically underprivileged neighbourhood, designers and researchers collaborated with residents to investigate how local community challenges could be supported by initially hypothetical autonomous urban robots that later became known as "hoodbots" (Jaskiewicz & Smit, 2024). The designers created an open-source kit of robotic parts without predetermined functions, which was used in a community centre to co-design and prototype small robotic vehicles. For instance, one robot was conceived by the citizens to facilitate recycling by collecting materials such as paper and plastics, yet its final purpose remained deliberately ambiguous.

The robot was showcased in the neighbourhood park by the residents who created it. Many passers-by paused, observed, and began to interact with the robot and its creators, pondering about the device's purpose. Some attributed it functions such as "a self-driving garbage bin", while others speculated about its potential capacity for other functions in the community, such as helping to integrate the community and share neighbourhood stories. It prompted many discussions on possible futures of the neighbourhood, and triggered the residents to collectively reimagine the roles robots could play in their community. Ideas included using the local bicycle workshop for robot maintenance or engaging the robot in a protest against neighbourhood gentrification. The intentionally open-ended design of the robot allowed these multiple interpretations and led to new prototypes developed in following workshops.

The experiment's success has been its ability to provoke unexpected interactions between designers, residents and the robots, in which new ideas and meanings arose. The open-ended iterative process of making, reflecting, and interacting with these prototypes became as significant as the artifacts themselves, fostering various forms of collective learning in the community, embedded in the complex social fabric of the city.

Figure 16.1. By-passers trying to make sense of the prototypes of the hoodbots, putting them to use in unexpected ways.

Design is often discussed as problem-solving, whereby "plans of action are devised to change an existing situation into a preferred one" (Simon, 1969). In this perspective, prototypes are representing solutions to a problem. Design can also be seen as a learning process, where designers learn through reflection-in- and reflection-on- actions that typically involve creating sketches, drawings and prototypes (Schön, 1983, 1993), where design artifacts and design cognition are mutually constitutive. In this perspective prototypes are, rather than solutions, tools for learning.

The hoodbots in the above example also served as such tools for learning: they generated ideas throughout the deployment process and its aftermath. Yet, unlike in a regular design process, here they supported the learning process of the local community, rather than of the designers alone. The learning cycle repeated multiple times and continues to this day, and allows the process to unfold in unexpected ways, beyond the control of, and later not even involving the designers whatsoever. Design cognition became radically distributed here as many various actors were involved. The complex co-creation processes spanned beyond

293

single creative sessions (Kwon et al., 2024) and became a "process of mutual learning and discovery, where new ways forward are generated collaboratively and where the 'ownership' of those ideas or directions is therefore shared" (Britton, 2017: p. 34).

In this chapter, we discuss such collective learning becomes enabled through design, when design cognition is distributed and "everybody designs" (Manzini, 2015), as is the case for social innovation. The question is: How might design contribute to collective learning, once many are involved? To answer this question, we leverage the concept of Fifth Order Design (Mortati, 2022) as a framework and blend it with Schön's reflection-in-action, while extending it with the recognition of involvement of many human and non-human actors (such as Hoodbots or animals).

Artifacts as Means for Collective Learning

Designers face much uncertainty as little is known, yet much needs to be decided on. They resort to abductive logic (Roozenburg, 1993; Dorst 2011), relying on creative leaps and "what-if" reasoning. Only through creating and reflecting on artifacts, such as sketches and models, they learn about the quality of their ideas. They iterate until they are content. "Idealtypes" serve to agree with stakeholders on aims (Stompff et al., 2024), prototypes serve to validate design hypotheses in practice. The proof of the pudding is in the eating: a logic that is rooted in Pragmatist philosophy (Dixon, 2020; Stompff et al. 2022).

There are various sorts of design artifacts that deliberately support learning. *Cultural probes* (Gaver et al., 1999) are packages of items such as maps, postcards, and other materials to "provoke inspirational responses" of users. *Technology probes* (Hutchinson et al., 2003) are "low-fi technology applications designed to collect information around use . . . provide inspiration for a new design space" (Boehner et al., 2007: p. 1078). However, technology probes rely on reactive participation (Graham & Rouncefield, 2008) and aim to support designers in learning how users respond to new technology, to "open up possibilities" (Boehner et al.,

2012). *Speculative design* artifacts are yet a different sort, geared for learning by the public. Speculative design is a critical design practice that explores ideas by means of possible futures through "what-if" scenarios and physical prototypes (Dunne & Raby, 2024, 2nd ed.). It invites the public to critically think about technologies, its consequences and aims to stimulate debate. Speculative design has been compared to thought experiments, enhancing public engagement (Barendregt & Vaage, 2021). Also, non-designers can be involved in imagining pluralistic futures in *participatory speculative design* (Farias et al., 2022), for example, to stimulate debate (Ye & Zhang, 2024). *Adversarial design* goes one step further and aims to provoke political discussions (DiSalvo, 2012).

At first glance, the hoodbots might be seen as conventional prototypes designed to field-test hypotheses. However, the robots had no specific goal and did not solve any particular problem. They could be framed as technology probes, enabling designers to better understand how citizens respond to robots. Yet, in this case, the citizens, not the designers, were the primary learners. The hoodbots might also be interpreted as ideal types, or artifacts of speculative or adversarial design. However, they did not represent possible futures, nor aimed to provoke the public, as the citizens themselves participated in their creation and demonstrated them in the neighbourhood in a straightforward manner.

Instead, the hoodbots case is best understood as facilitating a process of collective reflection-in-action. The robots were interpreted by diverse people in diverse ways, sparking ideas, experiments, and reflections—not primarily by the designers but by the community members. Thus, the hoodbots facilitated co-design not only in the sense of generating new concepts, but especially in the sense of shared, distributed knowledge generation.

Fifth Order Design

To explain the collective learning through the design process that we encountered in the above case, we use Mortati's concept of the Fifth Order Design (2022), extending the widely cited Four Orders of Design model of Buchanan (Buchanan, 1992). Mortati views Fifth Order Design as a transformative approach that prioritizes learning to support transitions.

Four Orders of Design

Buchanan's Four Orders of Design model describes the ends of design in relation to increasing levels of complexity:

- First Order Design focuses on signs, symbols, and images, aiming to communicate meaning.
- Second Order Design addresses physical objects, such as products and buildings, where designers focus on aesthetics, functionality, and manufacturability.
- Third Order Design involves actions, with designers creating meaningful experiences, such as in the development of services.
- Fourth Order Design pertains to systems, including organizations, regulations, laws, and the values that govern these systems (Buchanan, 2001).

Higher orders of design address increasing levels of complexity. While a graphic designer can independently create an effective logo, developing a mass-produced product requires a team of experts. Services introduce an additional layer of complexity, requiring coordination of activities over time. Designing at the systems level, however, encompasses the highest level of complexity, as it must account for the conflicting interests and values of multiple stakeholders.

Decisions made at higher design orders significantly influence lower-order design choices. For example, consider the printing industry (see also Stompff & Geraedts, 2008). In the business-to-business context, customers pay-per-print, with all costs for machines, consumables, and maintenance included. A Fourth Order design choice, determining how business is conducted, profoundly impacts many Second Order design choices. Engineers develop durable machines that are repeatedly refurbished, and parts of outdated models are repurposed in new products, requiring designers to create modern designs that effectively combine "old" and "new".

Figure 16.2. A Fourth Order design choice on how to conduct business heavily influenced lower order design choices. This seamlessly integrated product is a hybrid of old parts and new technology.

Fifth Order Design of Learning Environments

Recently, Mortati (2022) proposed a Fifth Order of design as "design is currently moving beyond the development of systems . . . where people relate to each other. It is devising learning systems in which new and different types of *agents* act" (ibid.: p.27, emphasis in original). She names this Fifth Order "Relationships", and explicitly includes *human* and *non-human* agents, such as artificial intelligence (AI), other species and micro-organisms. The goal of Fifth Order design is devising learning environments, focussing on designing the relationships with non-human actors who participate in a creative endeavour to empower collective learning. The challenge is to device learning systems that are not just attentive to human needs, but also to the environment and the planet.

297

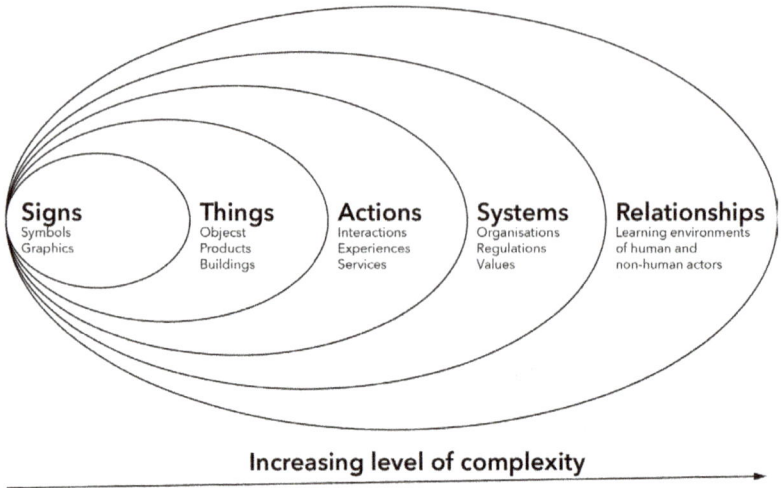

Signs
Symbols
Graphics

Things
Objecst
Products
Buildings

Actions
Interactions
Experiences
Services

Systems
Organisations
Regulations
Values

Relationships
Learning environments
of human and
non-human actors

Increasing level of complexity

Figure 16.3. The Five Orders of Design (Mortati, 2022). She added a Fifth Order to a well-known framework.

Fifth Order Design represents a significant conceptual leap. First, it downplays the contribution of humans as drivers of design cognition. With the rise of AI, it is increasingly evident that design and decision-making are no longer exclusively human domains (Giaccardi & Redström, 2020). Today, many choices are informed—or even determined—by black-box algorithms. Mortati further expands this perspective by recognizing "other species that live and thrive outside of human control" as potential design agents. While she introduces this as a foundational concept, she does not provide practical guidance. We interpret this as design whereby humans are no longer the sole contributors to the outcomes. For example, creating a shoe with mycelia involves a design process in which the final form of the shoe emerges organically during its creation (Figure 16.4).

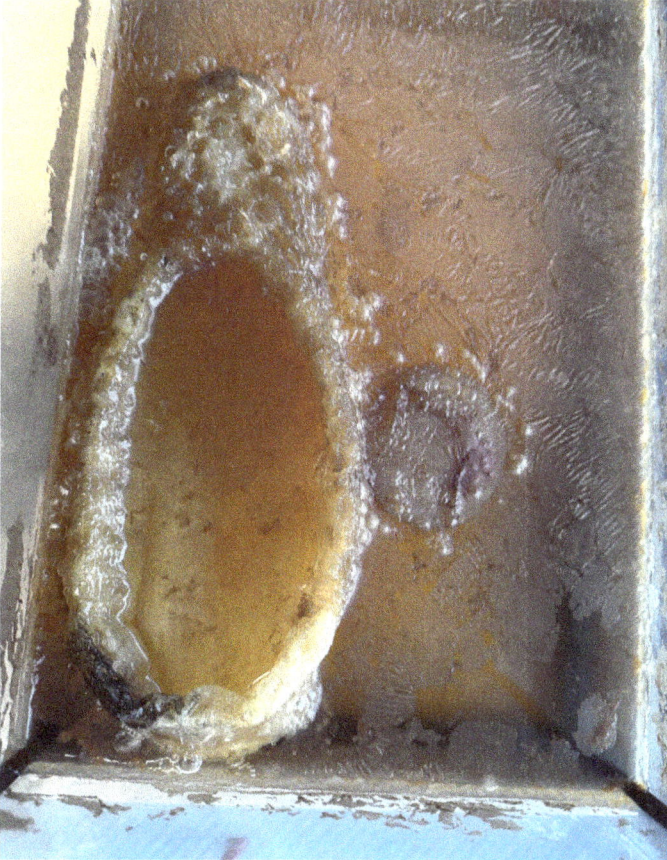

Figure 16.4. Designing a shoe with mycelia in a lab of the Technical University of Eindhoven.

The emphasis on the role of non-human actors in design resonates with Actor-Network Theory (ANT), a versatile framework that explores the evolving relationships between human and non-human actors (Latour, 1996). ANT has also been applied to design (Latour, 2008), placing the design of "things"—socio-material assemblies that self-organize around "matters of concern"—at its core (Storni, 2015). The equal consideration of human and non-human actors makes ANT particularly relevant for understanding social innovation (Britton, 2017, p. 67). Innovation is seen as the creation of new networks, actions, and artifacts. Visualizations, models, and prototypes are "non-human actors": integral parts of these fluid and interwoven networks.

Second, Fifth Order Design recognises collective learning through design as part of the process. Given the complexity and the involvement of multiple agents, there is no omniscient director guiding everyone. Instead, Mortati describes design as a continuous learning process, emphasizing data visualization and feedback loops (2022, pp. 28–31). The visual intelligence of designers, such as their capability to visualize data, should be leveraged for "visual problem seeking" (ibid., p. 33). Storni, whilst discussing ANT, also highlights the importance of visualization, particularly through the creation of "public maps" that are "constantly circulated, updated, collected, and made available to enable both engagement and reflection" (2015, p. 173). With such maps, actors can individually decide what to do and how to relate to others.

Fifth Order Design as Collective Reflection-in-Action

Visualization as a tool for reflection and learning brings Schön's reflection-in-action (1983; 1992) to the forefront. Reflection-in-action emphasizes that designers progress by reflecting on emerging outcomes. In this view, the design process and its artifacts can only be understood in relation to each other, just as design cognition can only be understood in relation to design practice (Kimbell, 2011). While Schön primarily focused on individual designers, subsequent studies examined monodisciplinary design teams (Valkenburg & Dorst, 1998) and large multidisciplinary teams (Stompff et al., 2016). Similar patterns emerged across these group settings, particularly in the reframing processes. These are pivotal moments when teams collectively make sense of unexpected events, interpreting the situation together and deciding on the next steps. For team learning, such surprises are the benefits (ibid.).

Fifth Order Design requires taking Schön's reflective practice to the level of society, where a diverse array of stakeholders is involved, including interest groups, citizens, policymakers, and the environment. In this context, issues lead to the emergence of "publics" in a Deweyan sense (DiSalvo, 2009; Dixon, 2018)—self-organizing groups that coalesce around shared concerns, engage in debate, and transform ideas into actions. Using the vocabulary of Actor-Network Theory (Latour, 2008), fluid networks arise as actors gather and transform "matters of fact" into "matters of concern." They contribute through action, by saying, doing or creating something that informs others, which prompts reflection and responses of others. This process inspires a continuous chain of action and reaction. Artifacts such as documents or maps crystallize during this process, shaping future relationships and plans. These artifacts, in turn, evolve as new contributions modify them. Within this dynamic web of actions, actors, and artifacts, new knowledge emerges (Mortati, 2022, p. 32).

Fifth Order Design is modelled here as a process of collective reflection-in-action of an ever-changing network of agents, shaping matters of concern and crystallizing into artifacts that represent knowledge and ideas. Both the artifacts produced and the network co-evolve.

Revisiting the Hoodbots as Fifth Order Design

The guiding question is: How can design contribute to collective learning when many individuals are involved? While Morati (2022) discusses data visualization and "visual problem seeking", the hoodbots represent another type of design artifact for collective learning. They are intentionally crafted as "semi-empty canvases" meant to be "painted" with meaning by others. They are mutable objects that allow individuals to engage, interact, and ascribe meaning. While not entirely empty—since they provide guidance on how to interpret them—they evoke different responses: for some, they are simply amusing objects, while for others, they serve as trash bins or street ambulances for the homeless (Jaskiewicz & Smit, 2024). Nevertheless, at face value, they are recognised as intelligent robots.

The hoodbots are intentionally crude and lack intricate details, communicating their unfinished, prototype status. They possess the "charm of the skeleton" (Stompff & Smulders, 2015), inviting others to engage in ways the designers had not anticipated and spark debate about what robots could imply for the community.

Robots serve as objects that facilitate the design processes of others, they are "design for design" (Ehn, 2008), enabling collective learning. Also, they allow us to articulate local matters of concern, such as waste issues, and foster the development of new relationships among various stakeholders.

Once designers intentionally create designs that promote collective learning, they engage in Fifth Order design. Their primary objective is not to produce graphics, products or services that meet the needs of others; rather, they focus on developing artifacts that facilitate collective learning and knowledge production, irrespective of the outcomes. As such, they may design maps, visualize data, or create artifacts such as robots, films, scenarios, or services that aim to inspire and motivate others. While DiSalvo (2012) and Storni (2015) argued that these artifacts should be agnostic, challenging, and provocative, the hoodbots demonstrate that such artifacts can also be inviting and inspirational. Through these creations, individuals can reflect, respond, take action, build new relationships, or choose to ignore them altogether.

To provide another example, students created a "Wishing Wall" for the neighborhood. The ambiguous nature of a "wish" provided insights into the various concerns present within the community. The wall literally functioned as a "semi-empty canvas" where meanings could be ascribed, facilitating collective sense-making. Designed to provoke thought and encourage reflection, the Wishing Wall sparked discussions that unfolded organically, without a predetermined end in mind.

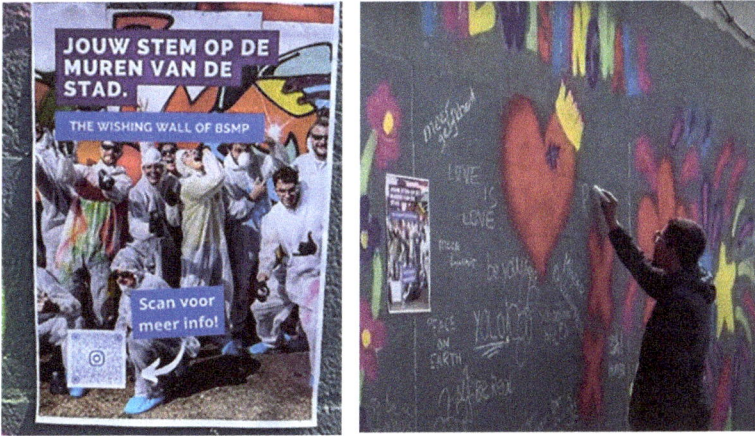

Figure 16.5. The Wishing wall in Amsterdam Noord where a passer-by is chalking down a wish.

Discussion: Design in the Real World

Papanek's influential book *Design for the Real World* (1971) challenged the dominant goals of design, focused on persuading consumers. He advocated for responsible human-centered products that take into account the planet's limited resources. Since its publication, the emphasis in design has shifted from consumer appeal to user needs, from aesthetics to usability, and from sales to circularity. Today, the core ideas also find their way into social innovation (Britton, 2017; Manzini, 2015) and transitions (Irwin, 2015; Tonkinwise, 2015).

However, the application of design for transitions received criticism (e.g., Mulder & Loorbach, 2018), for example, due to the promotion of "techno-fix approaches" (Richterich, 2023), leading to "solutionism" (Morozov, 2013). More important, design as a problem-solving strategy is fundamentally at odds with transitions. The challenges are genuinely wicked (Rittel & Webber, 1973), characterized by "long-term, complex, and non-linear processes of systemic change, which typically unfold over decades at the societal level. The high degree of ambiguity, unstructuredness, and uncertainty makes it exceedingly difficult to plan and design for transitions" (Boehnert et al., 2018: p. 894).

Transitions occur at the societal level, involving numerous agents— including non-human—who interact and influence the process. No single entity can oversee all, and it is a misconception to attribute design cognition solely to a specific group, as this presumes that they can fully anticipate the outcomes of their plans. The composition of this group, whether it consists of smart experts, or a temporary group engaged in co-design, does not negate the reality that their perspective is limited and influenced by their implicit value systems. This distinction between "designers" and "others," as well as the divide between "planning" and "executing" re-introduces the dualism between thinking and doing within the context of transitions. Instead, Mortati defines Fifth Order design as the creation of learning environments where a diverse array of agents can contribute. This approach promotes collective learning, eliminating the distinction between "designers" and "others".

In this chapter, we integrated Fifth Order design with Schön's concept of reflection-in-action. It envisioned design as a creative, exploratory, and deeply reflective form of collective learning where making and thinking are intertwined. The role of designers is to develop and create artifacts observable by many that enable collective meaning-making, producing "semi-empty canvases" that provide just enough structure to spark action while allowing others to "paint" and ascribe meaning. The goal of design shifts from crafting perfect solutions or aesthetically pleasing end-products to creating intentionally unfinished objects, maps, and scenarios that prompt ideas and reflections. This reflective process of thinking and doing, creating and reflecting, unfolds "within" the world and is inseparable from it. The issues that arise, the networks that develop, and the artifacts that are created all co-evolve and manifest as a single, interconnected "thing" (Latour, 2008).

Since Papanek's call to redefine the ends of design, many challenges (climate change, energy issues, and biodiversity loss) remain prevalent. Papanek envisioned designers as external agents capable of offering solutions "for" the world but underestimated the complexities intrinsic to these issues. However, designers and design researchers can still make meaningful contributions, by creating "semi-empty canvases" that are open for interpretation, malleable for change and incite ideas. These "canvases" facilitate collective learning, allowing for experimentation, the shaping of networks, and the emergence of temporary solutions that many can contribute to. We must transition from the mindset of "Designing *for* the Real World" to "Designing *in* the Real World," recognizing that we are all designers participating in these transitions.

Key Takeaways

No one can fully grasp the complexities of transitions. However, design can play a role—not by providing "solutions," but by enabling collective learning processes that involve many.

- Fifth Order Design goes beyond creating systems; it concerns learning systems where various agents, both human and non-human, interact.
- The catalyst for collective learning is reflection-in-action, through which fluid networks emerge around matters of concern, and knowledge solidifies into artifacts.
- Design supports collective learning, irrespective of the outcomes, by producing thought-provoking artifacts open to interpretation, allowing others to assign meaning.

References

- Barendregt, L., & Vaage, N. S. (2021). Speculative design as thought experiment. *She Ji: The Journal of Design, Economics, and Innovation, 7*(3), 374–402. https://doi.org/10.1016/j.sheji.2021.04.003

- Boehner, K., Vertesi, J., Sengers, P., & Dourish, P. (2007, April). How HCI interprets the probes. In *Proceedings of the SIGCHI Conference on Human Factors in Computing Systems* (pp. 1077–1086). Association for Computing Machinery. https://doi.org/10.1145/1240624.1240780

- Boehner, K., Gaver, W., & Boucher, A. (2012). Probes. In *Inventive Methods* (pp. 185–201). Routledge.

- Boehnert, J. (2018). *Design, ecology, politics: Towards the ecocene.* Bloomsbury Publishing.

- Boehnert, J., Lockton, D., & Mulder, I. (2018). Editorial: Designing for transitions. In C. Storni, K. Leahy, M. McMahon, P. Lloyd, & E. Bohemia (Eds.), *Design as a catalyst for change – DRS International Conference 2018* (pp. 1–9). Design Research Society.

- Britton, G. (2017). *Co-design and social innovation: Connections, tensions and opportunities.* Routledge.

- Buchanan, R. (1992). Wicked problems in design thinking. *Design Issues, 8*(2), 5–21. https://doi.org/10.2307/1511910

- Buchanan, R. (2001). Design research and the new learning. *Design Issues, 17*(4), 3–23. https://doi.org/10.1162/07479360152681156

- DiSalvo, C. (2009). Design and the construction of publics. *Design Issues, 25*(1), 48–63. https://doi.org/10.1162/desi.2009.25.1.48

- DiSalvo, C. (2012). *Adversarial design.* MIT Press.

- Dixon, B. (2018). From making things public to the design of creative democracy: Dewey's democratic vision and participatory design. *CoDesign, 16*(2), 97–110. https://doi.org/10.1080/15710882.2018.1483781

- Dixon, B. S. (2020). *Dewey and design.* Springer International Publishing.

- Dorst, K. (2011). The core of 'design thinking' and its application. *Design Studies, 32*(6), 521–532. https://doi.org/10.1016/j.destud.2011.07.001

- Dunne, A., & Raby, F. (2024). *Speculative everything: Design, fiction, and social dreaming* (2nd ed.). MIT Press.
- Ehn, P. (2008). Participation in design things. In *Participatory Design Conference (PDC)* (pp. 92–101). ACM Digital Library. https://doi.org/10.1145/2145351.2145362
- Farias, P. G., Bendor, R., & Van Eekelen, B. F. (2022, August). Social dreaming together: A critical exploration of participatory speculative design. In *Proceedings of the Participatory Design Conference 2022 - Volume 2* (pp. 147–154). https://doi.org/10.1145/3533000.3533081
- Friedman, K. (2000). Creating design knowledge: From research into practice. In *Proceedings of IDATER 2000: International Conference on Design and Technology Educational Research and Development*. Loughborough University. http://www.lboro.ac.uk/microsites/design/IDATER/2000/contributions/friedman/index.html
- Gaver, B., Dunne, T., & Pacenti, E. (1999). Design: Cultural probes. *Interactions, 6*(1), 21–29. https://doi.org/10.1145/320309.320316
- Giaccardi, E., & Redström, J. (2020). Technology and more-than-human design. *Design Issues, 36*(4), 33–44. https://doi.org/10.1162/desi_a_00504
- Graham, C., & Rouncefield, M. (2008). Probes and participation. In *Proceedings of the Tenth Anniversary Conference on Participatory Design 2008* (pp. 194–197). https://doi.org/10.1145/1386623.1386654
- Hutchinson, H., Mackay, W., Westerlund, B., Bederson, B. B., Druin, A., Plaisant, C., ... & Eiderbäck, B. (2003, April). Technology probes: Inspiring design for and with families. In *Proceedings of the SIGCHI Conference on Human Factors in Computing Systems* (pp. 17–24). Association for Computing Machinery. https://doi.org/10.1145/642611.642616
- Irwin, T. (2015). Transition design: A proposal for a new area of design practice, study, and research. *Design and Culture, 7*(2), 229–246.
- Kimbell, L. (2011). Rethinking design thinking: Part I. *Design and Culture, 3*(3), 285–306.

307

- Latour, B. (1996). On actor-network theory: A few clarifications. *Soziale Welt, 47*(4), 369–381.

- Latour, B. (2008). A cautious Prometheus? A few steps toward a philosophy of design: Keynote lecture. In F. Hackne, J. Glynne, & V. Minto (Eds.), *Proceedings of the 2008 Conference of the Design History Society* (pp. 2–10). Universal Publishers.

- Manzini, E. (2015). *Design, when everybody designs: An introduction to design for social innovation.* MIT Press.

- Morozov, E. (2013). *To save everything, click here: The folly of technological solutionism.* PublicAffairs.

- Mortati, M. (2022). New design knowledge and the fifth order of design. *Design Issues, 38*(4), 21–34.

- Papanek, V. (1971). *Design for the real world: Human ecology and social change.* Pantheon Books.

- Rittel, H. W. J., & Webber, M. M. (1973). Dilemmas in a general theory of planning. *Policy Sciences, 4*(2), 155–169.

- Schön, D. A. (1983). *The reflective practitioner: How professionals think in action.* Basic Books.

- Schön, D. A. (1992). Designing as reflective conversation with the materials of a design situation. *Research in Engineering Design, 3*(3), 131–147.

- Simon, H. A. (1969). *The sciences of the artificial.* MIT Press.

- Tonkinwise, C. (2015). Design for transitions – From and to what? *Design Philosophy Papers, 13*(1), 85–92. https://doi.org/10.2752/144871615X14223737776548

- Valkenburg, R., & Dorst, K. (1998). The reflective practice of design teams. *Design Studies, 19*(3), 249–271. https://doi.org/10.1016/S0142-694X(98)00004-5

- Ye, Y., & Zhang, D. (2024). Co-creating pluralistic futures: A systematic literature review on participatory speculative design. In C. Gray, E. Ciliotta Chehade, P. Hekkert, L. Forlano, P. Ciuccarelli, & P. Lloyd (Eds.), Proceedings of DRS2024: Boston, 23–28 June 2024 (Research Paper No. 1316). Design Research Society. https://doi.org/10.21606/drs.2024.1316

Guido Stompff, Tomasz Jaskiewicz, Troy Nachtigall, Iskander Smit

309

Mieke Oostra, Koen van Turnhout

17. An Acupuncture Approach for Societal Change: Reflections on the Make-ability of Society

DOI: 10.1201/9781003609766

Introduction

While the challenges and possible projects and topics you could contribute to in the energy transition of the built environment are enormous, the time, capacity and budgets available as a research group at the University of Applied Sciences to contribute to the current societal goals are limited. The research group Applied Urban Energy Transitions (AUET) feels the responsibility to max out these limited public and private resources for the betterment of society. How to select what to take on or not? Our purpose is to prove that it is possible to create new directions of solutions to address one or more societal challenges technically, economically and socially. However interesting and inspiring the results might be, it is not only the solution in itself that counts for us. The purpose of our projects is to change, boost and challenge thinking around the (im)possibilities in the energy transition together with (regional) partners. On the one hand, it is not (only) a solution with the aim to create a template or concept for others to work with, but also as a point of departure for change in the sector. On the other hand, it is an intervention with the intention to stir people, to break down mental barriers and to tickle and inspire professionals to join in and rethink what they can contribute to the energy transition. This chapter reflects on the approach chosen by AUET that focusses on the energy transition in the built environment.

Large-Scale vs Small-Scale Interventions

After World War II, large reconstruction plans were made to rebuild complete urban regions including city centres. Functionalism ruled, which meant that the prime focus was on the required functions, their capacities and the subsequent mobility and logistics required when planning and designing neighbourhoods, cities and regions. Community development, the wish to shape or reorganise your own environment over time (Habraken, 1961), the need for undefined space (Sennett, 2018), human scale (Alexander et al., 1977), social control (Jacobs, 1961) and environmental impact (Meadows et al., 1972) among others, were neglected. In the sixties and seventies this led to a counter-movement that demanded attention for the environment (Meadows et al., 1972), the human factor and small-scale development (e.g. Jacobsen,

312

1961; Schumacher, 1973; Alexander et al., 1977; Habraken, 1961). From these developments we have learned not only to address technological and economic challenges, but also social and environmental aspects. Information technology and the alternative organisational constructs that have become possible over time, have proven to be key enabling factors in facilitating large-scale interventions with the possibility for local and personal fine-tuning. We are currently in the midst of translating these technologies and insights into the context of retrofitting.

Change of Era

As a result of these learnings, in combination with a range of major societal challenges, we no longer experience a stable world with incremental change. The pace of innovation has increased as part of a transition of eras that requires step change (Rotmans, 2021). This means the possibilities for research are plentiful. How to decide what challenges to take on? In consequence, we have chosen to invest a substantial part of our capacity for imagination in the creation of artifacts to translate wanted possible futures into new concepts, props, processes, prototypes, archetypes, tools and pilot projects. Research, design, engineering, construction and testing are essential elements in order to be able to create artifacts that make the trespassing to this new world possible. And to do so as a collective, not only as individuals. The need for this is also recognised at the EU level with the New European Bauhaus program (EUI, 2022).

This corresponds to our own experience; it is not easy for people, even professionals, to form for themselves an image of how society may change, and what this means for the built environment and their own work or business. As designers and researchers, we have the capacity to create images and constructs that can help others in forming an image of the future, while it can also lead to valuable reflection and discussions about the proposed future. The more concrete, the bigger the effect we've noticed. It is not only our target to improve the quality of living and to substantially reduce carbon emissions locally by generating solutions for the retrofit of dwellings, but we aim as well to boost thinking and action directed at change of the retrofitting sector. Additionally, we hope to raise awareness that it is not only a question of retrofitting the building stock. Change of the building sector is needed, as is a fundamental change in how we look at buildings. Buildings not only need to be seen as physical objects that can be appreciated as homes and integral parts of the urban fabric. Buildings are currently transforming into actors in local developing energy systems. We aim with our work to boost the discussion of how all this may be realised. That's why we invest a substantial amount of our time, capacity and budget in realising prototypes, tools, or preferably, pilot projects and in creating infrastructure that can facilitate the wanted future to be born. AUET aims thus to clear barriers in thinking and contribute to the 'conception' of new ideas and solutions that will help realise the energy transition and through this improve the quality of life in urban regions. To be able to pinpoint this focus in our research the need arose for a new term: research acupuncture. With an acupuncture approach we suggest designers and researchers need to find sweet spots in the transition where a project with limited resources can have maximum effect. In this chapter we will describe: (1) the origins of the term, (2) an example and the context in which research acupuncture is applied by AUET, (3) outcomes, and we will end with (4) a reflection and discussion section.

The Origin of the Acupuncture Metaphor

Research acupuncture refers loosely to the term urban acupuncture, a phrase used in urbanism derived from traditional Chinese medicine. The urban acupuncture phrase can be used both as a metaphor or heuristic (Hemmingway & De Castro Mazarro, 2022). According to architecture critic and historian Kenneth Frampton (2000), the phrase was first coined by architect and urbanist Manuel de Solà-Morales. De Solà-Morales emphasised the importance of always intervening with 'concrete' matters while effecting change that goes beyond the physical (de Solà-Morales, 2008). Jaime Lerner, a Brazilian architect, urbanist, mayor and governor made the term known with his interventions and wrote a book dedicated to the topic (Lerner, 2014). Lerner himself was hoping to cure bad spots in the city with one little pinprick; an intervention in the form of a design assignment. According to Lerner many cities are sick or even terminal and have a need for an intervention that will release a blockage in energy flow to revive entire urban areas. Urban acupuncture therefore focusses on small-scale interventions with a rather large impact on the livability of neighbourhoods. The intervention can both be conducted by a professional or a bottom-up initiative. Also, other architects and urbanists used the acupuncture metaphor for their interventions; like Marco Casagrande (Casagrande, 2011), a Finish architect and social theorist.

Later the term eco-acupuncture was introduced to be able to refer to small-scale urban interventions aiming to facilitate pathways to a resilient low-carbon urban environment (Ryan, 2013) via design-research-engagement-action by Chris Ryan, professor Design and Sustainability at the University of Melbourne (Ryan, 2024). According to Ryan (2013) 'Eco-Acupuncture focuses on multiple small interventions in an existing urban precinct that can shift the community's ideas of what is permissible, desirable and possible and provide transformation points for a new trajectory of development to a resilient low-carbon future.'

From the mentioned literature on urban and eco-acupuncture the following key principles can be distilled: small-scale interventions, finding the right 'pressure points' or impact, citizen involvement, inspirational value and potential for system transformation.

Mieke Oostra started to use the research acupuncture metaphor to be able to describe interventions designed and engineered by her research group AUET and partners in order to explain that project ideas are picked to develop interventions with the potential to boost the energy transition in the built environment by creating new perspective both at local and systems level. For the research group the intention is to create a first sketch of the possible template and then elaborate on this as part of the development for a new 'regional ecosystem approach'. At the local level, the aim is to realise something concrete, something partners can use, and other people can see and touch in order to stir them, break down mental barriers and to let them understand both the approach and its

315

potential impact as a way to invite them to join in. The other aim is to create a point of departure for broader discussion in the region and beyond, to further thinking on what is needed to take it to the next level. What should be improved, added or changed to let it result in ecosystems that can systematically retrofit high-rises in the Netherlands? What barriers should be crossed and how? What kind of additional tools, processes and infrastructure is necessary in the form of tools, processes, regulations, instructions, investments in industrial production facilities, capacity development in workforce etc.? The intention is as well to tickle and inspire professionals to rethink what they can contribute to the energy transition and to fill in the blanks necessary to create a renovation wave.

An Application Example – Retrofitting High-Rises

One of the research lines of UAET is industrial retrofitting. Before the case of the retrofit of a specific high-rise is presented, we first discuss the context of the research acupuncture intervention.

The Context of Retrofitting Residential Buildings in The Netherlands

After World War II, contractors were challenged by the Dutch government to speed up production, not only to replace the residential buildings destroyed during World War II, but also to create the production capacity necessary to house the growing population in urban areas yet to be developed. Modernism not only introduced industrialised row housing and high-rise apartment buildings for the masses, but also led to an entire new way of urban planning and introduced new standards for living. Nowadays Dutch housing associations own nearly 30% of the dwellings, the highest percentage in Europe. This made the Netherlands a perfect place to start experimenting for an industrial approach to retrofitting. This seems to be a logical narrative now, but at that time it was not recognised as such.

To boost large-scale retrofitting was, and still is, a hard and slow process. For decennia an endless row of research and policy reports appeared, stating the importance of large-scale retrofitting in order to bring down fossil fuel consumption of dwellings. Also the realisation of a diverse set of pilot projects did not lead to the desired renovation wave. Approaches had to be tailor-made for every dwelling or building, and the budget required was to high for most private owner occupants, as well as to retrofit the entire real estate portfolios of housing associations. As a response to the difficulty to reach upscaling of retrofitting, the Ministry of the Interior and Kingdom Relations (BZK) financed Energiesprong. Energiesprong was a Dutch programme (2010-2016) aiming to upscale affordable ways of retrofitting. It is mainly known from its sub-program Stroomversnelling. The focus was put on row housing, since this is the dwelling type most common in the Netherlands. What was helpful is, that a substantial amount of Dutch row housing is owned by housing associations, professional parties, which reduces the complexity in decision making, since they can represent and bundle the interests of individual household groups.

Several schemes were tried: competitions, the coaching of private house owners, but both failed (Oostra, 2017). Therefore, a third schedule was designed to make a national deal that included 4 main contractors and 10 housing cooperations. The housing associations promised to bring in 111,000 houses that could be renovated block by block. The contractors promised to invest in an industrial production approach to bring down the costs per dwelling substantially. AUET participated in the adjacent projects, as well as some other research groups of universities of applied sciences. In this third scheme with the name Stroomversnelling, the Energiesprong succeeded in lowering prices per dwelling by 50%, to reduce time onsite by 90%, to provide a 10-year guarantee and; very important for social tenants; all at the same housing costs, which includes rent and energy costs. Additional deals were made at the provincial level. The Utrecht region was one of the first to work on a deal. The Stroomversnelling organisation was also located in Utrecht with a close link to our Utrecht University of Applied Science (UUAS).

317

Figure 17.1. The renovated Inside Out apartment building (photo: Rogier Lateveer).

The Choice to Retrofit a Specific Intervam High-Rise

After the Energieprong ended, all sorts of projects were ongoing and started to further develop the industrial retrofit approach for row housing. For AUET the focus was, and still lays on the further development of components and processes needed for the industrial retrofitting of row houses and the monitoring of their performance. To us the further development of integrated energy modules was very interesting, a new branch of the building installation sector that emerged as a result of the Stroomversnelling. Another research topic we took up was solutions for complex decision-making and customer journeys for both single households and owner associations. AUET realised together with some other regional partners that an industrial approach would possibly also suit other housing types. The retrofit of multi-story apartment buildings became the third research topic. First stock assessment surveys were made to investigate whether it would be possible to find a business case for an industrial retrofit approach for other housing types. The number of dwellings constructed in the sixties and seventies (55,000) executed with similar industrial construction systems, appeared large enough to make investments in dedicated production facilities possible.

Utrecht has several high-rises like the Brederode and Intervam apartment buildings in neighbourhoods like Kanaleneiland and Overvecht. The challenge seemed even more interesting since according to most, it was not possible to convert apartment buildings beyond six floors to energy neutral or positive.

Meanwhile, the Stroomversnelling continued to create the right preconditions for housing associations. To make investments in whole-house retrofits a more interesting alternative compared to the non-integrated mainstream approach, they put considerable effort in making sure all sorts of barriers were dismantled by proposing and lobbying for new national regulation, fiscal incentives, subsidies, standard building permit procedures at the municipality level, administrative commitment, and taking the Energiesprong approach abroad, thereby enlarging the market for industrial retrofit solutions. Partners in Utrecht decided to take on the challenge to figure out how to retrofit a specific Intervam apartment building with 10 floors could be conducted in a standardised and modular way as a first step towards component production on an industrial scale.

By taking on the challenge for a specific Intervam building a similar development for high-rise apartment buildings was started, as for row housing. Additionally, there were plenty of other reasons that fitted well with the aims of the research calls that came out at the time: to help businesses to create new energy solutions for buildings; to keep housing costs affordable for tenants despite home improvement and sustainable transport; to increase productivity in the renovation sector and to industrialise the retrofit approach for a new category of buildings. The most appealing reason, however, was that we thought 'this is cool!': since the general idea was that the conversion towards energy positive for the 10-story building was impossible, we had an idea. This idea entailed a technical solution that appealed to the imagination: the PowerNest, integrated roof modules that generate both solar and wind energy. In what would become the Inside Out proto-apartment project, the solution would be tried out for the first time. It was a challenge that appealed to our imagination.

319

An Application Example
– Outcomes

The general idea was that high-rise apartment buildings above 5 to 6 floors could not be retrofitted to the level of energy neutral since the building envelope is too small to mount all the necessary solar panels. This left energy positivity, the ability to generate more energy than needed in the building, beyond scope. The Inside-Out consortium managed to do the impossible; three one-year research projects financed by TKI Urban Energy resulted in the first energy-positive high-rise retrofit of 10 floors at the Henriettedreef in Overvecht. This provided the proof that an energy-positive high-rise retrofit was indeed possible. The pilot project made it realistic how solutions and components for a retrofit of an energy-positive high-rise could be integrated. Henriettedreef was a one-off, but it provided insight into how retrofitting of high-rises from the 1960s could be industrialised. This implicitly meant that the retrofitting of other high-rises constructed with the same or similar building systems could be approached in the same manner. The social tenants were generally very positive. Considerable effort had been made to involve them in the process, and it paid off.

The idea provided the opportunity to successfully apply for a TKI urban energy subsidy three times in a row. TKI urban energy started to use the outcomes of these Inside Out projects as an example for others. It has become an iconic project, both nationally and internationally. Professionals from across the globe still come to Utrecht to see this project with their own eyes. The attention resulted in a clear positioning of Utrecht as a region for the retrofit of high-rises, which was also picked up by the province. Subsequently this led to endorsement of the provincial deputy, causing Inside Out to be accepted as one of the developing regional ecosystems as part of a National Growth Fund project 'Toekomstbestendige Leefomgeving'. International interest led to an invitation to become part of the European Green Deal Project ARV. And back home in Utrecht, the Stroomversnelling had started to think about what preconditions were necessary for apartment buildings, and what this meant for the new regulations, incentives, etc. They were rigging up for streamlining the retrofitting of row housing.

Reflection & Discussion

This section will reflect on how the key principles of urban and eco-acupuncture are being met in the application example of Inside Out, research versus design in research acupuncture and the impact of some major changes in the context. The key principles of urban and eco-acupuncture as presented earlier were (1) small-scale interventions, (2) finding the right 'pressure points' or impact, (3) citizen involvement, (4) inspirational value and (5) potential for system transformation. We will start with these:

Small-scale interventions—The original concept of urban acupuncture refers to concrete interventions in the built environment to increase livability and to invite people to reoccupy and contribute to public space. The retrofit of the apartment building can be considered a small-scale intervention, since it only focused on one specific building, and the physical impact does not go beyond the borders of the neighbourhood, since some electric cars can be charged with the energy generated on the building. The opposite can also be argued: it is a large intervention since it does not limit itself to one apartment, but an entire building with 58 apartments with the intention to scale up to retrofit similar buildings.

Finding the right pressure points or impact—From hindsight the series of projects for the retrofit of apartment buildings might look like a very linear and planned process: you choose the 'pressure point' together with the idea, develop a (series of) project(s) and realise an impactful intervention that receives public attention. The reality is much humbler. The subsidy agency usually asks to define the problem, present a solution and make impact plausible (whether this is e.g. new knowledge, a practical solution or tool). The reality is that in times of transition, there is a complex situation in which many challenges and opportunities interconnect. As a

321

research group or consortium, you can monitor the developments and look for possible 'pressure points' that require interventions. Ideas will emerge that are likely to activate specific 'pressure points' and that can be clearly defined so they will fit within a stand-alone project. But ideas should also match the expertise, time, and enthusiasm available as well as with available calls for funding. Some ideas will never make it to the project stage. Some ideas do not receive enough enthusiasm in the research group or among partners, or other ideas may be more appealing or easier to raise funding for. Efforts to receive funding may not be successful. Some ideas make it to student projects instead. There is no guarantee that the idea will take off. So, the honest answer is that there is no way to recognise an opportunity to create something that has a much stronger impact than just being a solution to a concrete problem in a specific situation. There is, however, a way to increase the chance, and that is to question what the potential impact might be of certain ideas. Can impact be generated both at a local and systemic level? And what would that look like? With enough creativity in the research group or consortium, there should be no shortage of good ideas matching pressure points, from which the ideas with high potential can be picked. But there is no guarantee it will work out as planned.

Citizen involvement—Eco-acupuncture invites inhabitants to contribute with interventions in their own neighbourhood to diminish fossil fuel dependency and create new ways of living that will cause others to rethink their way of life, by developing alternatives and so inspiring co-citizens to create additional solutions that act as stepping stones towards a resilient low-carbon future. In reflection, it can be concluded that for this aspect, the research acupuncture approach is a disadvantage, since it is aiming for systemic change. This hampers the level of possible citizen involvement. They are involved, but the time available for interaction is limited. Without the ambition for systemic change, more time could have been spent on interaction with the tenants. From other projecs, we know that one-to-one contact and focus groups take a substantial amount of time. Then again, over the years, a lot of interaction with tenants in earlier projects has produced considerable insights on what aspects are important or even essential for tenants (e.g. Oostra & Nelis, 2022). The crux in manipulating this 'pressure point' is to relate insights to tenants' interests together with requirements from housing associations and retrofitting partners to take steps towards standardisation and modularisation. Then when looking for the other angle: renters' involvement was considered essential in the Inside Out projects. Compared to other, similar technical projects, the amount of time spent on interaction with tenants was even rather large. Social tenants were involved as co-designers in the project. As an example: in the test apartment project, a mock-up of the new façade was put up in the UUAS climate chamber, the space was furnished as an apartment and even a real size image of the apartment's view was put up behind the facade. Tenants were invited over to provide their thoughts and input. This resulted in useful input on maintenance and usage, which was then incorporated in further improvements.

Inspirational value—The attention the Inside Out projects generated among (international) professionals, and their questions and remarks, indicated that it shifted something in their thinking. The follow-up projects, directed at economic growth and upscaling, proved that others saw potential in the system change we are aiming for. Whether an intervention causes people to be inspired and to what extent can, however, not be predicted. AUET and its partners were aiming to influence the vision and approach of apartment building retrofits. The international attention was not something that was foreseen or even hoped for. It just happened to be so. What the concrete result is of the many professionals who came by to visit the project and what they do or change after the visit we don't know.

Potential for system transformation—Our aim with these projects was not to retrofit a specific building only. The idea was to start up the industrialisation process for an whole new typology of buildings. In this we succeeded. The projects can be regarded as a major shift in thinking on how to approach retrofitting of high-rises. There is now proof that a 10-story building can be made energy positive. More or less by accident, the outcomes of the projects also exemplified that buildings have a new role to play in

323

the local energy system. The change of the retrofitting sector will still require numerous additional steps. Several other professionals and subsidy bodies have confirmed the potential for systemic transformation. We are now in the process of trying to make this potential real.

Designers versus researchers—Looking at the consortium, the assignment was taken on both as researchers and as design professionals. Both used their ability for envisioning the future and translating this vision into concrete artifacts. And the division between researchers and designers is not black and white in the consortium since there were architects working as researchers and architects with a research mindset. The acupuncture metaphor seems therefore not only fit to indicate small local design interventions of architects, urbanists or bottom-up initiatives, but also the retrofit interventions presented in this chapter that were also aiming for change at a system level. But this is perhaps the essence of the acupuncture metaphor: creating great effect with one little pinprick.

Changes in the context—What we could not have predicted were the effects of external developments; the war in Ukraine for example, and its devastating effects on energy prices and the financial situation in general of the residents in Overvecht, a low-income neighbourhood. Due to inflation, costs for labour and building materials rose substantially. The war also influenced the supply chains of building materials. The supply chains were still dealing with the side effects from corona, but on top of that, security of supply fell due to the war. Due to these effects, discussions arose whether heavy investment in upgrading of the energy performance of the facades should be continued. Another important discussion dealt with the preferred solution: all electric or district heating. Inside Out was regarded as part of the all-electric solution toolbox. After the government had appointed municipalities as directors of the energy transition in neighbourhoods, district heating was presented as the preferred solution for the Overvecht neighbourhood of natural gas. Then due to continuing increase of local energy generation grid and electric mobility congestion became a hot topic in 2022–2023, and step by step also in the rest of the Netherlands. The general assumption was that a retrofitted apartment building with an all-electric solution would need a larger grid connection than before. That this is not necessarily the case, had to be proven in another research project. This meant, however, that net congestion required further development of the Inside Out concept and focus shifted from the entire Inside Out concept towards the energy module.

It can be concluded that steps towards the second aim for change at the sector level, after the retrofit of the Intervam apartment building, were heavily influenced by a changing context; (geo) politics, grid congestion, the lengthy process of legislative amendment for district heating and different considerations that were made in practice due to rising costs and technology development. They delayed the journey. It did not as such affect the direction of development, but it led to the necessary fine-tuning of the further development of the Inside Out modules. It is, however, a perfect illustration of the volatility of the context in which we all operate. So, if anyone in the consortium still thought the future could be planned, this person has by now experienced enough situations to wave the make-ability assumption farewell.

Key Takeaways

To conclude a few key takeaways as learnings from this chapter.

- As Rotmans has insightfully posed: there has been a shift from an era of change to a change of era. This means we are no longer operating in a world that can be considered as stable; therefore, we have to anticipate unexpected change to occur in the context in which we are researching and designing our interventions.
- The research acupuncture metaphor was introduced to make clear there is a multitude of project ideas we can choose from in an era of change. Why not take time to weigh the potential of several of them and choose ideas that have potential to create an impact beyond the situation they are created for?

325

- The series of projects presented also exemplified that the assumption of make-ability has passed its expiration date in many contexts in which practice-oriented research takes place. The concept of make-ability suggests that you can systematically plan the execution of an intervention as well as its impact. While this might be true in a stable environment with little change, in the current era of transition, this will only result in derailed planning and frustration. This does not mean you have no idea where an intervention might lead you and what following steps might be. But there is no way to predict what the exact outcome will be from pinpricking a chosen 'pressure point'. Also in the case of the apartment building, there was initially the idea that after a first retrofit the barrier for a similar retrofit would be lower. But due to changes in the context, this turned out not to be true. First the housing associations opted for district heating, then the plans suddenly stopped because of upcoming changes in regulation. Net congestion stifled all electric retrofit solutions, as did rising building costs. For this context applies that make-ability is something from the past. It is better to let go of the make-ability assumption and adopt a strategy of strategic navigation, a continuous analysis of bottlenecks and opportunities to determine which path is both interesting and feasible. During a research and design journey you are wise to constantly assess changes in the context and their potential effect on the goal you are aiming for, to anticipate change and to adjust the course whenever necessary.

References

- Alexander, C., Ishikawa, S., & Silverstein, M. (1977). *A pattern language: Towns, buildings, construction.* Oxford University Press.

- Casagrande, M. (2011, October). From urban acupuncture to the third generation city. *La Vie Magazine.* Malesherbes Publications. Retrieved from http://casagrandetext.blogspot.com/2011/10/urban-ecopuncture.html

- Frampton, K. (2000). Seven points for the millennium: An untimely manifesto. *The Journal of Architecture, 5*(1), 21–33. Taylor & Francis.

- Habraken, N. J. (1961). *De drager en de mensen: Het einde van de massawoningbouw.* Scheltema & Holkema. (English edition: *Supports: An alternative to mass housing,* 1972).

• Hemingway, J. M., & De Castro Mazarro, A. (2022). Pinning down urban acupuncture: From a planning practice to a sustainable urban transformation model? *Planning Theory & Practice, 23*(2), 305–309. Taylor & Francis.

• Jacobs, J. (1961/2016). *The death and life of great American cities.* Random House.

• Lerner, J. (2014). *Urban acupuncture: Celebrating pinpricks of change that enrich city life.* Island Press.

• Meadows, D. H., Meadows, D. L., Randers, J., & Behrens, W. W. (1972). *The limits to growth: A report for the Club of Rome's project on the predicament of mankind.* Universe Books.

• European Urban Initiative (EUI). (2022). *New European Bauhaus Compass.* Retrieved from https://www.urban-initiative.eu/sites/default/files/2022-12/NEB_Compass_V1.pdf

• Oostra, M., & Nelis, N. (2022). Concerns of owner-occupants in realising the aims of energy transition. *Urban Planning, 7*(2), 45–57. Cogitatio Press.

• Oostra, M. (2017). Experiences from de Stroomversnelling: nZEB renovation of row housing. Presentation held for construction partners of VTT, Helsinki.

• Oostra, M. (2015). De-burden or co-design & co-create? The future of open building. *CIB W104 Conference*, September 9–11. ETH Zürich. Retrieved from https://www.research-collection.ethz.ch/handle/20.500.11850/110955

• Schumacher, E. F. (1973/2011). *Small is beautiful: A study of economics as if people mattered.* Ballantine Books.

• Sennett, R. (2018). *Building and dwelling: Ethics for the city.* Allen Lane, Penguin Books.

• Rotmans, J. (2021). *Omarm de chaos.* De Geus.

• Ryan, C. (2024). About eco-acupuncture. Retrieved from https://msd.unimelb.edu.au/research/projects/completed/VEIL/eco-acupuncture/about

• Ryan, C. (2013). Eco-acupuncture: Designing and facilitating pathways for urban transformation, for a resilient low-carbon future. *Journal of Cleaner Production, 50*, 189–199. Elsevier.

Angelique Ruiter

18. Design Is Not the Answer! Reflective Practices and the Social Field: A Case Study in Amsterdam Gaasperdam

DOI: 10.1201/9781003609766

Introduction

Gaasperdam, located in Amsterdam's southeastern region, is a diverse neighborhood with a population of approximately 30, 000 residents. The area is characterized by a mix of social housing and privately owned homes, with several residential high-rises built during the 1970s and 1980s. In terms of education, the neighborhood exhibits lower educational attainment levels compared to the city average, with a significant portion of residents holding only secondary education degrees. Unemployment rates in Gaasperdam are relatively high, often exceeding the average rates seen across Amsterdam, highlighting socio-economic challenges within the community.

In terms of energy transition, Gaasperdam is part of broader efforts within Amsterdam to move towards sustainability, with initiatives focusing on energy efficiency and renewable energy integration. The municipality aims to improve energy infrastructure and involve local communities in adopting sustainable practices. However, challenges such as socio-economic and educational disparities, and varying levels of community engagement often complicate these efforts, reflecting a broader need for inclusive and equitable energy transition strategies across urban areas.

The Green Hub Gaasperdam project Gebiedsgericht SLIM decided to address these complex dynamics of social and just energy transitions within the Gaasperdam community. The local organization has been situated in Gaasperdam, Holendrecht since 2019 and has been working on creating a healthy and sustainable neighborhood. The municipalities' subsidy Kansen voor West (Assets for Amsterdam West) provided an opportunity to bring together the Green Hub, the municipality, local entrepreneurs, energy companies, and Utrecht University of Applied Sciences, to work on the question: how can we facilitate a local energy transition that is social, democratic and just, within a 2-year-long research study. The study was to focus on three aspects; how to improve the socio-economical position of the residents, how to align city-wide developments about the energy transition with empowering residents to develop their neighborhood, and how to make steps towards making houses go natural gas-free.

Theoretical framework

Design and Social Change

Design as a practice aimed at designing things has recently been transforming into a practice that helps to address complex societal transitions (Bijl-Brouwer, 2020). A solution-oriented approach, as is the case with using design principles, often doesn't consider the complex nature of transitions and the societal context. Furthermore, the innovative approach of design can counteract sustainability goals if it is not embedded in social change (Mulder, 2007). Looking at how social change is created, we need to acknowledge that social systems dictate the reality of everyday life and it is these social systems that cannot fully be captured by designers (Vink, 2024). Furthermore, we need to acknowledge that the social systems we take into account are also connected to the natural world as they are intertwined (Dewey, 1980.) To aim at changing such social systems, one would have to study and interact with the daily routines of our embodied experience in the social settings in which we move to map the social structures (Mead, 1972) that need addressing. In this light, the role of the designer may not be to design something physical, but rather to facilitate ways in which the communities that form social systems can be strengthened and better connected.

Living with and Emergence

Part of facilitating this lies in adapting to the complex and changing circumstances in which we work. Wakkery's concept of "living with" (2018, 2023) is helpful here, as he underscores the ever-evolving and adaptive relationship between individuals and their environment. Wakkery's framework highlights how design interventions should remain open to the fluid and dynamic nature of people's engagement and social interactions, reflecting the ongoing negotiation between people, their environment and technology, and the changing context over time. Furthermore, Gaver's concept of emergence (Gaver et al., 2022) also aligns with the unpredictable, spontaneous behaviors of people and insights that arise as people interact with their environment and the social environment in ways that were not initially anticipated. In such a seemingly volatile environment, this begs the question of where we can begin to look for ways to make sense of such unpredictable social interactions.

Social Structures as a Switch for Change

As it happens, research has shown that changing existing social structures is the most effective way to create the systems change (Meadows, 2008) that designers, or transition facilitators, seek to accomplish by addressing complex societal transitions. Here it is pivotal to make the invisible social structures visible. We can map out the fabric of the social structure in all its facets, such as norms, values, and even performativity (Butler, 1990) to see what the social structures and routines in a certain context look like. Rather than using approaches such as Giga mapping, that may place the system outside of our experience (Vink et al., 2021), we have to use approaches that are aimed at reflection and transformation. This means we have to find methods to allow the systems to see itself and methods that allow us to stay present with what is and feel into new possibilities (Scharmer, 2009).

The Social Field

To tap into social structures and become aware of the social fabric and routines in a context it is vital to use methods that allow us to experience this social construct. In doing so it is even more vital that we can understand what happens in a social context that comes together in a certain space at a certain point in time; in other words, we need to understand the *particular* social interaction of people who come together to work (in whatever way) on societal transitions. It is here that social fields come in. A social field can be viewed as: "The entirety of the social system with an emphasis on the source conditions that give rise to patterns of thinking, conversing, and organizing, which in turn produce practical results" (Scharmer, Pomeroy & Kaufer, 2021, p. 5). In other words; the social field is that which makes up the quality of how we are together and how that determines our thoughts, actions, and any outcomes that follow out of our meeting. You can think of this quality as follows; we have all been in meetings where the atmosphere was tense and we also have been in meetings where we have felt inspired.

Both situations influenced how we felt, how we thought, and how we acted. Being able to turn a social field from low quality, e.g. tense, where people are fearful, to high quality, e.g. where people are inspired and get creative (Pomeroy and Herrmann, 2023), is a game changer in working with the challenges of our time. Scharmer and Pomeroy (2024) have suggested that being able to help the world out of its current crises comes down to helping the social field evolve through fourth-person knowing: a state in which individuals and collectives can, a.o., sense the whole *and* sense the possibilities.

How to Facilitate Social Field Knowing

How do we bring people to this state, and how will it lead to results? It seems designers, or transition facilitators, have work to do. Systemic design will, e.g., have to look beyond the system and take into account the systemic design within, as the quality of the intervention depends on the interior condition of the intervener (Scharmer, quoted in Huffington Post). So if we want our designs and solutions to be in line with what is asked in the world, we need to be reflective of our thoughts, feelings, and actions. Furthermore, we need to be able to connect to ourselves and others. In the step of facilitating we need to come up with ways to help individuals and communities to reflect upon themselves, we can use methods such as journalling, or social presencing theater. We also need to develop skills in deep listening and sensing to know how to respond to what is happening in groups and to act from our embodiment, rather than our cognitive perception. We need to be able to hold space. Last but not least, we need to step into the social fields ourselves if we want to learn and understand. And if we sense that we have sensed the possibilities, it is then that we can start prototyping and perhaps come up with ways to interact with the world and others differently and work towards more sustainable communities (Hummels, 2023).

333

Case Study: Gaasperdam

Challenges

As communities become key actors in the energy transition, the involvement of residents is crucial. However, this requires overcoming significant barriers, including skepticism and distrust of local government, which are common in such initiatives (Seyfang & Haxeltine, 2012; Van der Schoor & Scholtens, 2015). Our research reflected a broader trend in community-based energy projects where daily struggles often overshadow long-term goals (Bauwens et al., 2016). Distrust of the local government is another problem frequently noted in the literature on participatory energy governance (Walker & Devine-Wright, 2008). The cause of this low trust in the municipality is due to historic differences where the municipality had made decisions without consulting residents. This complicated efforts to engage the community. Internal project challenges, such as the lack of project management capacity and turnover in personnel, further exacerbated these issues.

Complex Stakeholder Landscape

The complexities of stakeholder dynamics between residents, the Green Hub, local entrepreneurs, the municipality, and energy experts such as Energie Samen and Escom, and community feedback in the context of the energy transition are shaped by a combination of socio-economic factors, trust issues, educational disparities, and diverse expectations. Addressing these complexities requires a careful, tailored approach that emphasizes inclusivity, clear communication, and equitable distribution of benefits. By using participatory methods and engaging stakeholders at all levels, the Gaasperdam project was able to foster moving towards a just and sustainable energy transition that meets the needs of all residents.

The main challenges were Affordability vs. Sustainability: while energy efficiency and renewable energy solutions may be framed as cost-saving measures in the long term, upfront costs or perceived inconvenience discouraged participation. Residents in Gaasperdam who were struggling financially viewed thinking about the energy transition as an additional burden rather than a potential benefit. Another problem was Social Equity: Ensuring that the benefits of renewable energy and energy efficiency projects are equitably distributed (i.e., not just benefiting wealthier, privately-owned households but also low-income residents in social housing) is essential. If the transition is perceived as furthering inequalities, put upon residents by the municipality, it breeds resentment and opposition. Another challenge we faced when working with the stakeholders was trust issues: Residents harbored skepticism towards the government due to prior experiences with projects that have not addressed residents' immediate needs or concerns. This resulted in resistance to new initiatives, even those aimed at improving sustainability and quality of life.

Engagement of the Community

Other challenges concerned the engagement of the community: It required significant effort to build trust with residents, especially as part of the residents felt marginalized and were skeptical of top-down interventions by researchers. Furthermore, it was a challenge to work with power imbalances: even within a participatory framework, there are still underlying power imbalances. Green Hub staff members with a higher level of education and a different social background sometimes dominated discussions, sidelining the voices of less vocal residents. Additionally, local governance structures prioritized certain viewpoints over others. There were Different Levels of Engagement: Not all community members felt equipped or willing to engage deeply in the participatory process.

Methodology

Creating an Experimental Space

Initially, our research was grounded in an Asset-Based Community Development (ABCD) approach, which proved useful for mapping the community's existing skills, assets, and needs. However, we quickly encountered challenges in engaging individuals who could commit time and energy to the further development of the community. This was worsened by the fact that the Green Hub staff was already stretched thin with daily operational tasks, and residents, understandably, had other priorities. Employing a living lab approach, or creating space for an experimental space, allowed us to create a setting where we could a setting in which we could apply deep listening and phenomenological approaches, instead of working from rigid design plans. Furthermore, it allowed us to empower residents and stakeholders to contribute actively (participatory design) to the process of working on a shared understanding of the energy transition in Gaasperdam. This approach was vital for being able to capture emergent themes that lived within the community.

To further adapt to these challenges, we employed ethnographic observation alongside a phenomenology approach, which allowed both residents and staff to articulate their lived experiences of connecting with the neighborhood, the Green Hub, sustainability, and personal development. Through this approach, we were able to remain responsive to the fluctuating needs of the community, adjusting our research and interventions accordingly. This flexibility enabled us to co-design solutions that supported both resident growth and the acceleration of the energy transition. Furthermore, the use of these methods allowed us to observe how the social field within the Green Hub was evolving over time, providing insights into the ongoing transformation of the community and the project as a whole.

Interventions

For the present research, this concept was operationalized through a series of interventions aimed at fostering engagement and co-creation among the Green Hub staff, residents, and other community stakeholders. These interventions included:

- The co-creation of a skill-share map, in which residents identified their skills and the skills they sought from others to promote community development and knowledge exchange. The resulting list of skills was distributed to all participants, and a poster displaying the full skill inventory was prominently displayed within the Green Hub. This tangible representation of skills aimed to encourage collaboration and foster a sense of community engagement.

- Development of a Learning Agenda: In collaboration with Green Hub staff and residents, a five-year learning agenda was established to identify and prioritize the goals for the community's future development.
- Reflection Circles: Engaging residents, students, and Green Hub staff, reflection circles provided an opportunity for participants to explore their experiences of living in Gaasperdam. These sessions aimed to "feel into" the social field by employing interactive methods such as games and facilitated discussions about participants' thoughts, feelings, hopes, and fears regarding the neighborhood's livability. This process created a feedback loop, enabling community members to share insights and contribute to a collective understanding of the social environment (Scharmer, 2009).
- Co-creation of Communication Materials: Residents and Green Hub staff collaborated in designing communication materials to raise awareness about the Green Hub's existence and activities. This also included the development of a game that helped participants understand the principles of the doughnut economy and how they could engage with one another to initiate skill-building activities, fostering a deeper understanding of sustainability in the local context (Raworth, 2017).
- Mission Mapping Sessions: In partnership with staff, residents, and municipal representatives, mission mapping sessions were held to reflect on the past two years of the project. This process allowed stakeholders to acknowledge both successes and areas for improvement. Additionally, the sessions examined the stakeholder constellation and discussed potential changes or additions to the network, culminating in a co-creation of future actions aligned with the needs and goals of the community (Scharmer, 2009).

Outcome

Because the project lasted for 2 years, researchers were able to observe how residents' and the Green Hub's attitudes toward the energy transition changed over time. Phenomenological interviews and ethnographic observation allowed for the capacity to observe what was emerging in the Green Hub. Instead of the fixed routes of ABCD community development, and the outcome it is working towards, we found that to start building anything in Gaasperdam, approaches had to start with building up individuals. We also saw that individuals who became more confident and started to feel more ownership of the neighborhood drew in more people to the Green Hub. Residents told us they felt happy and at home within the Green Hub setting and that it was a place where they felt motivated to work on things.

In this respect, we focused on interventions that would support an ongoing awareness of sustainability and individual needs on a local level. We did so by organizing sessions where we reflected on what residents and the Green Hub's staff were occupied with, what they needed, and how those needs might (eventually be tight to the goals of Kansen voor West (Assets for Amsterdam West) to speed up the energy transition in Gaasperdam. Through our interventions, we were able to map the psychological needs surrounding the energy transition in Gaasperdam. We observed the residents' trajectory of coming into the Green Hub and connecting to themselves and their surroundings. This made for growing self-worth and broadened the social network of the residents. From having a feeling of enhanced self-worth, residents would start to take part in activities of the Green Hub, which made for skill building and a greater interest in and feeling of ownership in the neighborhood. Having gained more confidence and having built up a social network and skills that might lead to paid work in the future, residents found more space to think about matters concerning sustainability. Some started directing their activities to e.g. the local food garden in the Gardens of Brasa, a community garden, or by taking part in the Hub's green waste program. This, in the end, made residents more open to thinking about energy transition solutions.

This was also true for the material surroundings where we saw that residents started to feel more connected to the community around the Green Hub and started working on, for example, keeping the grounds of the Gardens of Brasa, which resulted in the municipality giving residents rights to maintain the land. Some residents also started working on the Green Hub's bio-digester, which would provide energy at a local level.

338

Through the methods, e.g. learning agendas, mission mapping, and reflection circles, the Green Hub staff, in co-creation with residents, was able to self-prioritize the goals for the coming year. Furthermore, reflection on the past 2 years brought forth questions about the capacity for project management and possible sources of funding. The municipality told us they had learned valuable lessons on how to approach community-driven initiatives. The more activist public officials saw that approaches based on empowering individual residents would work best and were ready to advocate for them in their organizations. Reflection circles and mission mapping also created a space for addressing trust and power imbalances between the Green Hub residents and the municipality. Long-standing feelings of power imbalance were, naturally, not solved by these interventions, but they did make all stakeholders feel better connected and better understood.

Finally, co-creating a game that educated residents on the doughnut economy proved to be a good conversation starter when inviting new residents and local entrepreneurs to the Green Hub. Furthermore, it proved to be a way of reflecting on what people wanted to accomplish, as well as on how to do so in a way that would contribute to a more sustainable Gaasperdam. The co-creation of communication materials to reach more residents in Gaasperdam proved less successful. The materials were piloted in the neighborhood but did not bring the number of new residents we hoped for.

Discussion and Conclusion

By centering reflective practices and social field analysis, this research demonstrates the value of a more emergent, responsive, and community-driven approach to facilitating participatory processes around the energy transition. Traditional design-led models may overlook important social and relational dimensions, whereas the approach used in this study empowered residents to meaningfully participate and co-shape the ideas that would allow for a social and just transition, in line with their own needs and aspirations.

339

Improving the social field by improving individual well-being, seems to be key to creating an environment in which people are more open to thinking about sustainability and acting towards a more sustainable neighborhood. The extent to which this would lead to concrete willingness and/or action towards a natural-gas-free Gaasperdam remains to be seen. As for public officials, the improved social field also seemed to motivate them to want to create change, in the way the municipality goes about funding local initiatives, within their organization. In this sense, the hypothesis of Pomeroy and Scharmer, that a better-quality social field inspires better actions, seems to take hold. For this study, it also held true that improving the social field seemed to yield far better results than a design-led approach would have done. Design can, though, contribute in a positive way to foster an improved social field by facilitating deep listening and reflection (e.g. through the doughnut deal game).

While the reflective and social field-oriented approach proved valuable, it also brought several challenges. For example, it was difficult to get residents completely engaged due to competing priorities. Additionally, the contextual nature of the research makes it challenging to generalize the findings to other communities. Further complicating the replication of the results is the fact that this approach takes time. Researchers will need to engage with the context, live with the context, to learn to understand how to respond to emerging patterns and social changes.

This research highlights several important areas for future research. First, there is a need to further develop methods and frameworks for incorporating reflective practices and social field analysis into participatory processes in the energy transition. Second, longitudinal studies are needed to understand how the social field evolves and the long-term impacts of this approach.

Key Takeaways

- Working directly towards creating a just energy transition did not seem the right approach. Rather, the first focus should be on a shared vision of the energy transition and/or sustainability, where residents define priorities for social justice, democratic governance, and sustainability, and other stakeholders (e.g. the municipality and energy companies) look for a balance in facilitating these priorities. This could serve as a roadmap for future actions and inspire other initiatives working on the same goals (e.g. Escom used a similar individual empowerment approach in The Hague).

- Identifying power imbalances that affect democratic engagement in the energy transition helps to create an understanding of how decision-making structures (e.g., local government) may be set up in such a way not only to exclude marginalized voices but also may reflect (to residents) past colonial structures. This helps come up with first insights into how to overcome these barriers through inclusive governance through empowerment strategies and community input.

- Working with reflection circles created a more closely-knit social field: it allowed residents to discuss ideas and plans in an open and safe space; this helped to bridge gaps between diverse stakeholders. Residents also felt more motivated to adopt sustainable practices or participate in future initiatives through further development of the connection with the neighborhood. Furthermore, all stakeholders understood better what each other's concerns and motivations were.

- When working with processes around sustainability and the energy transition, it is key to get a nuanced understanding of the psychological and emotional dimensions of the transition. This means focusing on the lived experience of the stakeholders involved to capture an in-depth understanding of social and psychological needs. This could inform strategies to overcome individual and collective barriers to adopting sustainable energy practices.

341

- Co-developing and co-designing communication materials, and gamified solutions that encourage participation and engagement, e.g. with the doughnut economy and sustainability, fostered greater local ownership of creating a more sustainable Gaasperdam. It seems that creating a tangible product together can help facilitate a feeling of having one's voice heard.
- Ongoing engagement and creating a long-term strategy for improvement is difficult. Staff changes at the Green Hub and discontinued subsidies made it hard to create continuity for the local community. It would make sense to ensure that at least the quality of the social field and its impact would be monitored over a longer period.

References

- Bauwens, T., Gotchev, B., & Holstenkamp, L. (2016). What drives the development of community energy in Europe? The case of wind power cooperatives. *Energy Research & Social Science, 13*, 136–147.

- van der Bijl-Brouwer, M., & Malcolm, B. (2020). Systemic design principles in social innovation: A study of expert practices and design rationales. *She Ji: The Journal of Design, Economics, and Innovation, 6*(3), 386–407.

- Butler, J. (1990). *Gender trouble: Feminism and the subversion of identity*. Routledge.

- Chambers, R. (2008). *Revolutions in development inquiry*. Routledge.

- Cunha, M. P., & Chia, R. (2007). Using teams to avoid peripheral blindness. *Journal of Organizational Change Management, 20*(4), 460–471.

- Currano, R. M., Steinert, M., & Leifer, L. J. (2011). Characterizing reflective practice in design – What about those ideas you get in the shower? In S. J. Culley, B. J. Hicks, T. C. McAloone, T. J. Howard, & P. Badke-Schaub (Eds.), DS 68-7: Proceedings of the 18th International Conference on Engineering Design (ICED 11), Vol. 7: Human Behaviour in Design, Lyngby/Copenhagen, Denmark, 15–19 August 2011 (pp. 374–383). The Design Society.

- Dewey, J. (1934). *Art as experience.* Perigee Books.

- Gaver, W., Krogh, P. G., Boucher, A., & Chatting, D. (2022). Emergence as a feature of practice-based design research. *Proceedings of the 2022 ACM Designing Interactive Systems Conference.* https://doi.org/10.1145/3532106.3533524

- Hargreaves, T., Hielscher, S., Seyfang, G., & Smith, A. (2013). Grassroots innovations in community energy: The role of intermediaries in niche development. *Global Environmental Change, 23*(5), 868–880.

- Hummels, C., Trotto, A., Peeters, J., & Levy, P. (2018). Design research and innovation framework for transformative practices. In *Strategies for Change.* Glasgow Caledonian University.

343

• Jones, P. (2018). Contexts of co-creation: Designing with system stakeholders. In P. Jones & K. Kijima (Eds.), *Systemic design* (Vol. 8, pp. 1–18). Springer. https://doi.org/10.1007/978-4-431-55639-8_1

• Kretzmann, J. P., & McKnight, J. L. (1993). *Building communities from the inside out: A path toward finding and mobilizing a community's assets.* ACTA Publications.

• Mead, G. H. (1972). *Mind, self, and society: From the standpoint of a social behaviourist.* University of Chicago Press.

• Meadows, D. H. (2008). *Thinking in systems: A primer.* Earthscan Publications.

• Mulder, K. (2007). Innovation for sustainable development: From environmental design to transition management. *Sustainability Science, 2*(2), 253–263. https://doi.org/10.1007/s11625-007-0036-7

• Pomeroy, E., Herrmann, L., Jung, S., Laenens, E., Pastorini, L., & Ruiter, A. (2021). Action research from a social field perspective. *Journal of Awareness-Based Systems Change, 1*(1).

• Raworth, K. (2017). *Doughnut economics: Seven ways to think like a 21st-century economist.* Chelsea Green Publishing.

• Reason, P., & Bradbury, H. (2001). *Handbook of action research: Participative inquiry and practice.* SAGE Publications.

• Scharmer, O. (2014, February 5). Collective mindfulness: The leader's new work. *Huffington Post.* Retrieved from https://www.huffpost.com/entry/collective-mindfulness-th_b_4732429

• Scharmer, O. (2016). *Theory U: Leading from the future as it emerges* (2nd ed.). Berrett-Koehler Publishers.

• Scharmer, O., Pomeroy, E., & Kaufer, K. (2021). Awareness-based action research: Making systems sense and see themselves. In D. Burns, J. Howard, & S. M. Ospina (Eds.), *The SAGE handbook of participatory research and enquiry.* Sage Publications.

• Schön, D. A. (2016). *The reflective practitioner: How professionals think in action.* Routledge.

• Seyfang, G., & Haxeltine, A. (2012). Growing grassroots innovations: Exploring the role of community-based initiatives in governing sustainable energy transitions. *Environment and Planning C: Government and Policy, 30*(3), 381–400.

- Spinuzzi, C. (2005). The methodology of participatory design. *Technical Communication, 52*(2), 163–174.
- Tomico, O., Wakkary, R., & Andersen, K. (2023). Living-with and designing-with plants. *Interactions, 30*(1), 30–34.
- Vink, J. (2023). Embodied, everyday systemic design – A pragmatist perspective. *Design Issues, 39*(4), 35–48.
- Vink, J., Wetter-Edman, K., & Koskela-Huotari, K. (2021). Designerly approaches for catalyzing change in social systems: A social structures approach. *She Ji: The Journal of Design, Economics, and Innovation, 7*(2), 242–261.
- van der Schoor, T., & Scholtens, B. (2015). Power to the people: Local community initiatives and the transition to sustainable energy. *Renewable and Sustainable Energy Reviews, 43*, 666–675.
- Wakkary, R., Oogjes, D., Lin, H. W. J., & Hauser, S. (2018). Philosophers living with the Tilting Bowl. In CHI 2018 – Extended Abstracts of the 2018 CHI Conference on Human Factors in Computing Systems: engage with CHI (Article 94). Association for Computing Machinery. https://doi.org/10.1145/3173574.3173668
- Walker, G., & Devine-Wright, P. (2008). Community renewable energy: What should it mean? *Energy Policy, 36*(2), 497–500.

Danielle Arets, Bart Wernaart,
Jeroen de Vos, Rens van der Vorst

19. Crafting Conflict: Designing Meaningful Interactions in a Smooth Society

DOI: 10.1201/9781003609766

Introduction

In recent years, numerous cities have embraced the concept of the so-called Smart City to address urban challenges through the implementation of smart technology (Kitchin, 2019). Utilizing devices such as smart cameras, sensors, and data technology, these cities aim to enhance various aspects of urban life. Additionally, alternative terms such as 'ubiquitous,' 'digital,' or 'u' cities have been employed to describe similar concepts (Mechant et al., 2012).

A common denominator is that smart cities look at technology as a solution to complex urban issues and societal challenges like climate change, energy transition, poverty, inclusivity or safety (Pali & Schuilenburg, 2020) and offer a playground to experiment with smart city solutions (Van Veen & Visser-Knijf, 2022). There is often an unbridled optimism about technology solving these complex issues (Townsend, 2013). This can be related to the strong push of tech companies that put a 'tech'-focus at the forefront of the urban agenda (Alizadeh, 2017).

There is, however, growing criticism of this solutionist approach (Morozov & Bria, 2018). It results not only in wrong and narrow-minded assumptions around the Smart City concept but also in a political narrative that is "nonideological, commonsensical, and pragmatic" (Kitchin & Dodge, 2014, p. 131). Scholars warn of the risk that smart cities turn into technocratic forms of governance (Datta, 2015), focusing on prediction and solutions and that governing becomes calculating (Pali & Schuilenburg, 2020).

Furthermore, the blurring of the relationship between public and private partners is stressed. Over the past 15 years, more and more companies have become heavily involved in the public sphere. As a result, there is a risk that business interests will take precedence over social and cultural interests and turn cities into a Smart City market.

On top of that, there are growing concerns about data ownership and privacy. There is a lack of political and legal clarity about to whom the data belongs. Publicly generated data are often passed on to private companies without supervision (Marvin &

Luque-Ayala, 2017). This raises ethical issues concerning privacy surveillance that have significant consequences on how citizens are conceived and treated. There is a risk—especially because of a severe lack of public oversight (Naafs, 2018)—that data profiles could be utilized to categorize areas for targeted policy interventions or marketing efforts. In general, we could say that these ethical issues tend to be overlooked, and are not at the core of the design of the smart city (Bietti, 2022; Van Veen & Wernaart, 2022).

What possibly worries scholars the most is that citizens are progressively being dehumanized, portrayed either as 'parcels of data' (Morozov, 2014) or as mere 'data points' (Kitchin, 2013). There is a growing concern that individuals are viewed primarily as a means for developing smart solutions rather than as residents living in a city (Laurent & Tironi, 2015). Chourabi asserts that humans are often overlooked in the Smart City discourse, as technological and policy aspects tend to dominate (Chourabi, 2012).

Van Zoonen furthermore stresses that citizens are not only overlooked, but often dismissed as their concerns or critique is flagged as non-sensical (Van Zoonen, 2020). Angry citizens protesting smart lamp posts or 5G technology, are not infrequently flagged as ill-informed, and their concerns are discarded as irrational (Van Zoonen, 2020).

However, many citizens, not only the ones protesting, appear to have ambivalent feelings about the Smart City. There appear to be 'tipping points' during discussions on smart city developments, whereby an initially positive attitude towards the use of sensor technology, such as increasing safety, changes during the discussion (Snijders et al., 2019, p. 90).

As to Van Zoonen, the public value approach to Smart Cities is rooted in a deliberative perspective of democracy with a dominant rational discourse for problem-solving: the idea that a rational exchange of ideas should result in the best decision possible. However, this deliberative approach falls short when responding to conflicts or uncomfortable feelings (Van Zoonen, 2020; De Droog-Arets, 2024).

Designing Friction through Adversarial Design

There is a growing attention to the need to highlight dissent and (design) friction in relation to the developing smart cities (Baibarac-Duignam, De Waal, 2021; Rasch, 2020; De Droog-Arets, 2024). Smart Cities are eliminating friction (Rasch, 2020). After all, there are no traffic jams and there is no air pollution. With that, there is also no room for conflicts.

The German philosopher Byung-Chul Han sees the pursuit of 'das glatte' [smoothness] (Han, 2015) as a sign of a zeitgeist in which we are striving for a slippery society with the use of digital technology. With our current focus on efficiency, we increasingly avoid difficult situations. Smart devices help us to seemingly navigate around busy traffic situations, ensuring that slow-moving seniors do not obstruct our speed, but with that, we lose our ability to deal with friction and uncomfortable situations. Instead of designing a smart application, we should design elements of friction that help to move beyond the seamless design dogma of our smart cities and engage citizens in a meaningful debate on the desirability of these developments.

As to Chantal Mouffe, the fact that dissatisfaction, friction, and feelings of resistance or conflicts around Smart City developments are hardly considered, indicates, a broader societal problem that applies not only to the theme of the Smart City, but more generally to our dealings with conflict. Currently, we fail to recognize and deal with conflicts and dissent in society. Angry citizens, illustrate that there is a great public dissatisfaction with the political structures in which people are "allowed to vote but have no voice" (Mouffe, 2013, p. 117).

Mouffe regards agonism, which involves acknowledging differences, as a necessary element for a thriving democracy. According to her, in our existing system, prevailing policies and values are imposed upon us. This suggests that certain groups of citizens will inevitably find themselves at odds with these policies, leading to feelings of exclusion and underrepresentation, ultimately resulting in frustrations and conflicts (Mouffe, 2013).

Mouffe emphasizes the importance of highlighting diverse perspectives and believes that public debates play a crucial role in achieving this. However, she argues that current debates need to be restructured because they neglect the concept of agonism. She suggests that our society debates should focus on embracing the multitude of opinions, varied values, and emotions surrounding societal issues, and underscores the significant role that artists and designers can play in this as they can highlight conflicting values and emotions at stake (Mouffe, 2013).

Adversarial Design to Highlight Dissent

DiSalvo (2015), applying Mouffe's philosophy to the practice of design, argues for adversarial design to do the work of agonism through design. Through designed products, artefacts, and experiences, we can create space for productive dissensus and contestation (DiSalvo, 2015, p. 125). Adversarial designs can give form to discussions, making them less vague and confusing but also assist in the process of skilled examination and reconstruction, rendering problematic situations perceptible.

Adversarial designs come in many shapes and forms, but most of them tend to highlight elements of frictions and discomfort in our Smart Society as well as enhancing a debate on the desirability of smart city developments. An illustrative example is the *Controversing Data* project by Baibarac-Duignan and De Waal (2021). In their research on data visualizations, they come to understand that slick, aesthetically pleasing data representations leave little room for critical reflection on the data. By turning these data visualisations into controversial ones, they noticed individuals can become engaged (Baibarac-Duignan and De Waal, 2021, p.1).

Designing controversy or friction seems counter-intuitive to the design discipline; after all, designers are trained as problem solvers. However, over the past decades, there has been growing attention to design that is critical instead of affirmative (Dunne & Raby, 2024).

There is a thriving practice of designers and scholars focusing on the role of design in the reconfiguration of policy (Bason, 2016, Kimbell, 2016), and the re-imagination of democratic systems (Macini, 2015) as well as redesigning our public debates (DiSavlo, 2015; De Droog-Arets, 2024).

As for DiSalvo, the goal of adversarial design is not necessarily immediate change; instead, it aims to stimulate debate and function as tangible evidence in political discourse (DiSalvo, 2015, p. 13). According to DiSalvo, adversarial design serves as both an inquiry and a practice.

As it aids in bringing clarity to complex situations, enabling individuals to perceive, experience, and act upon them.

Design Practice: The Moral Design Game

To explore how adversarial design can support a smart city debate, in which friction has a place, we designed a Moral Design Game, a board game (Figure 19.1) in which participants (maximum 8 per game round) take on various societal roles: the alderman, an employee of a data company, a human rights activist, a journalist, a concerned citizen, etc.

Figure 19.1. Moral Design Game.

The central element of the board game is the value framework of Schwartz et al. (1990; 1992; 1994; 2012; 2014) (Figure 19.2), and the communication philosophy of Jürgen Habermas (1990). The main goal is to trigger a moral dialogue amongst the participants with the aim to explore possible (preferably exhaustively) moral viewpoints, learn what the friction between these viewpoints look like and explore design principles that can be used to overcome this.

 The participants contribute to the dialogue based on two main principles (conform Habermas): 1) radical equality and 2) with access to the same information. For this, to erase existing hierarchical/unequal professional and social relationships between participants, the participants are given a role that is dissimilar to their own professional or social position and have to 'act' in line with this role (Boes et al., 2007).

All participants are given the same information regarding a
moral smart city case. The discussion is held alongside the value
framework of Schwartz et al. (2014), which includes four quadrants
of opposing value sets: openness to change, self-transcendence,
self-enhancement, and conservation. The quadrants openness
to change and self-transcendence represent values that are
free of fear. The quadrants self-enhancement and conservation
represent values that are fear avoidant. The quadrants openness
to change and self-enhancement have an individual focus, where
the quadrants self-transcendence and conservation have a societal
focus. In the game, the pawns are used to reflect the position of
the participants (as they play their character), which results in
the visualization of the value friction that results from the moral
dialogue. The players are challenged to overcome this value friction
by introducing design solutions.

Figure 19.2. Value Framework.

The gameplay is as follows: at the start of the game, the players watch a short video in which a scenario, or a 'dilemma' concerning the smart city, is explained to them. For instance, the desirability of drones delivering parcels in the city center, being on one hand sustainable and fast, but on the other hand, causing noise disturbance, and furthermore, new commercial bodies are interfering in the public infrastructure.

After watching the scenario, the participants, guided by a moderator, will start the discussion. They do this by speaking from their assigned role about values that are important in weighing whether to allow drones in this case. The ultimate challenge of the game is to reach a consensus that balances these competing values.

During a first Moral Game adversarial design activity at NEMO Science museum (Figure 19.3) in Amsterdam (November, 2021), two researchers were present to observe and take notes on how the conversation unfolded in behaviors and body language. We noticed that this adversarial design activity prompts players to contemplate the predominant values guiding decision-making among diverse stakeholders in the city. As one player mentioned: *"I found it very engaging to play the alderman, because I suddenly got a real sense of the responsibilities that come with that."*

Figure 19.3. Game play session, NEMO, Amsterdam.

The game furthermore highlights instances where certain values may be at odds with each other; for example, the alderman prioritizing responsibility versus the NGO emphasizing privacy. This inner multiplicity evokes multiple ways of looking at smart city developments (Asenbaum & Hanusch, 2021) and empathizing with different worldviews (Kolto-Rivera, 2004).

As one player described it aptly: *"It is difficult but very fascinating to stand in someone else's shoes. It gave me the courage to speak out much more than I normally would have. I was allowed to be an entrepreneur, which made me very proactive in advocating for a better digital infrastructure. As a citizen, however, my position is different. By experiencing this friction, I learned a lot about the complexity of multi-stakeholder management."*

Furthermore, we noticed that the tangible nature of the game board offers a concrete vessel to discuss developments that are predominantly 'in the cloud' or driven by 'black box' technology (De Droog-Arets, 2024) which makes these developments comprehensible. *"This activity has also taught us that playing a game helps us to discuss the complex theme of the Smart City in a concrete way."*

Discussion

To enrich public debates on Smart Cities, it is essential to move beyond traditional methods like focus groups, dialogues, and consensus meetings. Callon emphasizes that hybrid forums should incorporate creative approaches that prioritize listening, learning, and community building, even when these processes involve discomfort. Like Mouffe (2015) and Han (2015), Callon highlights the importance of addressing friction and diverse perspectives. These hybrid spaces are not merely about discussing "a world already made" but about collaboratively shaping "a world in the making" (Callon, 2006).

In this context, we found that the Moral Design Dilemma game fosters more tangible and meaningful debates on smart city developments. Beyond facilitating discussion, the game proves to be an effective research tool for exploring societal frictions associated with new technological applications. It also serves as a playful platform for collaboratively designing solutions to these challenges.

The game's tangible board design helps translate often abstract and technical discussions into concrete terms, making the conversation more accessible. Additionally, its role-playing elements bring differences to the forefront, encouraging mutual discovery and understanding. The adversarial nature of the board game, where participants engage with moral dilemmas, allows them to experience power dynamics and discuss value conflicts in a more nuanced and engaging way.

We believe that this practice can grow further by conducting more targeted research on how adversarial design can be meaningfully embedded in (research) practices aimed at enriching the debate on the smart city.

Key Takeaways

The use of adversarial design, in this case in the form of a designed board game, within the context of the smart city can be a meaningful way to question the desirability of smart city developments. The lessons we have learned in this regard are:

- The complex discussion about the smart city, where much of the technology is invisible and not easily grasped by everyone, is addressed in detail through the use of a game board. The concrete designed game board helps make this discussion practical and tangible.
- The role-playing aspect, where players are asked to take on the position of, for example, a council member, an SME, or an activist, provides participants with a good understanding of the value conflicts involved in this discussion.
- Finally, this research teaches us that the use of adversarial design can be a meaningful way to investigate this topic, as through the gameplay meaningful knowledge is created.

References

• Alizadeh, T. (2017). An investigation of IBM's Smarter Cities challenge. *Cities, 63*(1), 70–80.

• Asenbaum, H., & Hanusch, F. (2021). (De)futuring democracy: Labs, playgrounds, and ateliers as democratic innovations. *Futures, 134*, 102836.

• Baibarac-Duignan, C., & de Lange, M. (2021). Controversing the datafied smart city: Conceptualising a 'making-controversial' approach to civic engagement. *Big Data & Society, 8*(2), 20539517211025557.

• Bietti, E. (2020). From ethics washing to ethics bashing: A view on tech ethics from within moral philosophy. In Proceedings of the 2020 Conference on Fairness, Accountability, and Transparency (FAT '20)* (pp. 210–219). ACM. https://doi.org/10.1145/3351095.3372860

• Boess, S., Saakes, D., & Hummels, C. (2007). When is role playing really experiential? Case studies. *Proceedings of the 1st International Conference on Tangible and Embedded Interaction*, 279–282. https://doi.org/10.1145/1226969.1227025

• Chourabi, H., Nam, T., Walker, S., Gil-Garcia, J. R., Mellouli, S., Nahon, K., ... & Scholl, H. J. (2012, January). Understanding smart cities: An integrative framework. In *2012 45th Hawaii International Conference on System Sciences*(pp. 2289–2297). IEEE.

• Datta, A. (2015). New urban utopias of postcolonial India: 'Entrepreneurial urbanisation' in Dholera Smart City, Gujarat. *Dialogues in Human Geography, 5*(1), 3–22.

• De Droog, D. J. A. M. (2024). *Save the debate: Through adversarial design*. University of Technology Eindhoven.

• DiSalvo, C. (2015). *Adversarial design*. MIT Press.

• Dunne, A., & Raby, F. (2024). *Speculative everything, with a new preface by the authors: Design, fiction, and social dreaming*. MIT Press.

• Habermas, J. (1990). *Moral consciousness and communicative action*. MIT Press.

• Han, B. C. (2015). *Die Errettung des Schönen*. S. Fischer Verlag.

359

• Kitchin, R., & Dodge, M. (2014). *Code/space: Software and everyday life.* MIT Press.

• Kitchin, R. (2019). *The ethics of smart cities.* Maynooth University.

• Laurent, B., & Tironi, M. (2015). A field test and its displacements: Accounting for an experimental mode of industrial innovation. *CoDesign, 11*(3–4), 208–221.

• Marvin, S., & Luque-Ayala, A. (2017). Urban operating systems: Diagramming the city. *International Journal of Urban and Regional Research, 41*(1), 84–103.

• Mechant, P., Stevens, I., Evens, T., & Verdegem, P. (2012). E-deliberation 2.0 for smart cities: A critical assessment of two 'idea generation' cases. *International Journal of Electronic Governance, 5*(1), 82–98.

• Morozov, E. (2014). *To save everything, click here: The folly of technological solutionism. Journal of Information Policy, 4*, 173–175.

• Morozov, E., & Bria, F. (2018). *Rethinking the smart city: Democratising urban technology.* Rosa Luxemburg Stiftung.

• Naafs, S. (2018, March 1). 'Living laboratories': The Dutch cities amassing data on oblivious residents. *The Guardian.* https://www.theguardian.com/cities/2018/mar/01/smart-cities-data-privacy-eindhoven-utrecht

• Pali, B., & Schuilenburg, M. (2020). Fear and fantasy in the smart city. *Critical Criminology, 28*(4), 775–788. https://doi.org/10.1007/s10612-019-09447-7

• Rasch, M. (2020). *Frictie.* Bezige Bij bv, Uitgeverij De.

• Schwartz, S. H., & Bilsky, W. (1990). Toward a theory of the universal content and structure of values: Extensions and cross cultural replications. *Journal of Personality and Social Psychology, 58*(5), 878–891. https://doi.org/10.1037/0022-3514.58.5.878

• Schwartz, S. H. (1992). Universals in the content and structure of values: Theoretical advances and empirical tests in 20 countries. *Advances in Experimental Social Psychology, 25*, 1–65. https://doi.org/10.1016/S0065-2601(08)60281-6

• Schwartz, S. H. (1994). Are there universal aspects in the structure and contents of human values? *Journal of Social Issues, 50*, 19–45. https://doi.org/10.1111/j.1540-4560.1994.tb01196.x

- Schwartz, S. H., Cieciuch, J., Vecchione, M., Davidov, E., Fischer, R., Beierlein, C., Ramos, A., Verkasalo, M., Lönnqvist, J.-E., Demirutku, K., Dirilen-Gumus, O., & Konty, M. (2012). Refining the theory of basic individual values. *Journal of Personality and Social Psychology, 103*, 663–688. https://doi.org/10.1037/a0029393

- Schwartz, S. H. (2014). Functional theories of human values: Comment on Gouveia, Milfont, and Guerra (2014). *Personality and Individual Differences, 68*, 247–249. https://doi.org/10.1016/J.PAID.2014.03.024

- Snijders, D., Biesiot, M., Munnichs, G., & van Est, R., with van Ool, S., & Akse, R. (2019). *Burgers en sensoren – Acht spelregels voor de inzet van sensoren voor veiligheid en leefbaarheid.* Den Haag: Rathenau Instituut.

- Townsend, A. M. (2013). *Smart cities: Big data, civic hackers, and the quest for a new utopia.* W. W. Norton.

- Van Veen, M., & Visser-Knijff, P. (2022). Ethics in local politics: A case study of the city of Eindhoven. In B. Wernaart (Ed.), *Moral design and technology* (pp. 149–168). Wageningen Academic.

- Van Veen, M., & Wernaart, B. (2022). Building a techno-moral city: Reconciling public values, the ethical city committee, and citizens' moral gut feeling in techno-moral decision-making by local governments. In Proceedings of the OpenLivingLab Days Conference 2022: "The city as a Lab, but now for real!" (pp. 115–128). European Network of Living Labs (ENoLL).

- Van Zoonen, L. (2020). Publieke waarden of publiek conflict: Democratische grondslagen voor de slimme stad. *Justitiële Verkenningen, 46*(3).

Risk Hazekamp,
Synechocystis sp. PCC 6803

20. "You Press the Button, We Do the Rest" – How to Cede Agency to the More-Than-Human?

DOI: 10.1201/9781003609766

Introduction

The ability to make is innately embedded in art and design practices. Makers create or produce something. At the same time, 'make-ability' is a complex word. It flirts with what critical theorist and filmmaker Elizabeth A. Povinelli (2016a, 2021) calls "late liberalism", while it also carries a hint of the sustainable self-reliance of transcendentalist philosopher Ralph Waldo Emerson (1841). The following text addresses the notion of 'make-ability' in the realm of bioart and biodesign by focusing on the more-than-human collaboration within my Professional Doctorate research project, entitled 'Unlearning Photography: Listening to Cyanobacteria'. This text also draws on the 'make-ability' in the rise and fall of the Eastman Kodak Company and the intimate—and hitherto seemingly inextricable—link between photography and toxicity. With the aim to face photography's toxicity head-on, I explore my own personal journey as an artist working with the medium of analogue photography: from spending months at a time in the photographic darkroom to no longer being able to enter that same space without getting nagging headaches. This venomous ferocity became the catalyst to upend my artistic practice and investigate possibilities for regenerative change and (self)care, resulting in a (re)search towards 'living photographic processes'. Through a combination of autoethnographic writing, embracing *unlearning* and *uncertainty* as methodologies, and two extensive experimental case studies from my own research project, I will unfold the quest so far. In my research, I engage with *Synechocystis* sp. PCC 6803 as a more-than-human 'Other' who potentially has access to knowledge towards a living photograph. Inviting this unknown 'Other' inside my research, or even more so, ceding agency to them, requires not only cessation of reasoning in problem and solution, but also consideration of practice-based art and design research as constantly in flux, as an intersectional figure of radical emergence (Gaver et al., 2022).

Impossible-to-Answer-Questions

A day in mid-August, one of those days with a heat warning. I leave early in the morning towards a forest I have been visiting a few times a year throughout my life. The seasons change the paths from narrow overgrown summer forest trails to deep tractor ruts when the forester drives the tracks open again in winter. In my head, I visualise the route I want to follow and make choices: along the Old Oak, via the sheep, the wild boar path and then to the hunting lodge back along the brooklet via the meadow. All the tracks and all the side paths are familiar; I have walked them all back and forth over and over and still, I can get lost.

This time, everything seems different. The path over which I always descend from the hill on which our wooden cabin stands is suddenly owned by a family that is going to build their new house in a seemingly impossible spot. A little further on, the production forest next to the grand Red Beech has been cut down. Everything not a deciduous tree is harvested and piled up like matchsticks. The resin confuses my brain; this wonderful smell is the wound fluid of the trees that now lie here. A forest reduced to cubic metres of material. Next, the road to the Old Oak has been transformed into a two-lane mud track, scraped by a massive shovel, one of those machines that leaves huge abrasions on pine, oak, birch, ash and cherry trees that have lived here for decades as autotrophic creatures via photosynthesis from the sun. I touch each of them and say I am sorry.

I know these tracks; at least, I know the roads constructed by man. But there are so many more pathways, and even the ones made by humans are constantly changing. Tracks change according to where wildlife or water is, where trees are to be cut down or other forest work is done. Suddenly there is a road where there never was a road. Suddenly, a road is gone where there always was a road.

Absorbing or dismissing knowledge, collaborations, growing expertise, relationalities, all contribute to the way I move within my artistic practice, where many of the roads I explore echo those described above. Yet in my more recent research, suddenly there are new byways and winding paths, completely unknown to me. For five years, I have been caring for what within "dominant science" (Liboiron, 2021a, p. 20n77) is referred to as *Synechocystis* sp. PCC 6803, a model organism I did not know existed. *Synechocystis* are Cyanobacteria, a phylum of organisms invisible to the human eye, but to whom we owe our existence. This life form changed the Earth's atmosphere by releasing large quantities of oxygen as a leftover of the photosynthetic reaction they set in motion during the Great Oxidation Event some 2.3 billion years ago (Aiyer, 2022). Little by little, Cyanobacteria shaped the evolution of respiration and made Earth inhabitable for humans.

'Make-ability' becomes a rather complicated endeavour when contrasted with what Cyanobacteria have 'produced' on this planet. When *Synechocystis* entered my conscious life, I had no idea how to care for this unicellular organism: what do they need, what level of care is appropriate? It is evident that these concerns should be commonplace when designing with the living (Zhou et al., 2024). Philosopher María Puig de la Bellacasa (2017, p. 24) writes: "From the perspective of human–nonhuman relations in technoscience and naturecultures, unproblematic visions of care (…) would not only be meaningless but could be fatal." As a researcher, as a human being, can I collaborate with a microbe outside my own body? Am I able to truly descend from the hierarchical evolutionary pyramid and feel equal to *Synechocystis*, a single-celled prokaryote? One aspect that would answer these questions right away with 'no' is that our idea of both time and space is on a completely different spectrum. Where *Synechocystis* moves around in geological deep time with their interconnected omnipresence, I live in minutes and days and years and move my body mostly between what I call home and other places I frequently visit. This difference between microbial and human time and space makes any attempt to come together a ludicrously upfront failed exercise. And yet, this is where my research is set. In the words of marine microbiologist Karen Lloyd (2022): "I try to think in a longer view and I don't feel pressure to have a utilitarian use for what I am finding right now. (...) If we all say I'm not going to explore this unless it is going to make someone more money or save a life or make somebody happier in an immediate sense, then we are going to miss out on so much (...). Instead, I try to find the biggest most ridiculous questions, like impossible-to-answer-questions for me that I feel I have some expertise on, and then try to meet it in the middle, try to come up with some way to get a little bit closer to it."

My personal expertise lies with analogue photography. Almost as long as I have been criss-crossing that one patch of forest, I have spent time in the photographic darkroom. I watched my father's ever-shaking hands transfer photos from one tray to another and was fascinated by the red/yellow glow we found ourselves in. What stuck most was the scent in this small room where images were created by conjuring with light. The smell of the chemicals before 'odourless fixing bath' existed, the smell that for me will always be linked to magic. Since I've been reading Elizabeth A. Povinelli (2016b, p. 26), I understand that also I cannot "simply kick [my] habit of associating the astringent signature smell of photographic chemicals" with the moments seeing photographic images come alive. Povinelli writes: "The sensory history of chemicals sear into the affects, creating bonds of desire, nostalgia, and mourning for the very toxins now slowly overheating bodies and landscapes."

Kodak Moments

It is a confusing activity, walking in familiar territory. You think you know it and yet you continually stray. The points I recognise are often linked to smell, like the places where hunters lure and kill game, each and every hunting season. The mixture of blood, pheromones and mud never fails to linger. The places where killing took place are often overgrown again by brambles or ferns or wild strawberries, but the smell of mischief remains.

Everything changes continuously and everything remains the same, matter comes, goes and yet stays equal. So does energy, a constant that only takes on a different form. "Here, the first law of thermodynamics, commonly known as the law of

*'conservation of energy', becomes poetics, a reassuring mantra:
energy and matter cannot be created or destroyed. It is of such
beauty and comfort to know that the amount of energy and
matter remains the same. Everything is constantly becoming
something else" (Hazekamp, 2023, p. 22).*

"You press the button, we do the rest." was the famous advertising slogan of the Eastman Kodak Company, commonly referred to as Kodak (Fineman, 2004). The company dominated the world of analogue film material, both in photography and in the motion picture industry. Its iconic motto was first used in 1888 and Kodak held on to this phrase until its bankruptcy in 2012. The slogan emphasizes the capturing of every special 'Kodak moment' in life by just a click of the camera shutter, only to jump over the entire photographic development process looking straight towards the end result: the photographic images. This can be seen as a clear "device paradigm" (Borgmann, 1984; Robbins et al., 2016) of modernity's consumer capitalism, where often process is something tedious best outsourced to low-income countries. Digital image editing services located in 'places of otherness' have provided the Global North with lightning-fast delivery of clean retouched photographic images.

Over the last decade, the term 'Kodak moment' evolved into something completely different. It became a metaphor for a company's inability to adapt, for a business that misses the opportunities of the moment and in doing so fails to keep up with the changes of an industry it helped build (Anthony, 2016). And the Eastman Kodak Company certainly did this, for while making sure people didn't spend any time thinking about the process behind the creation of photographic images, Kodak developed its own downfall. Because, although it was a Kodak engineer who invented the first digital camera in 1975, still "Kodak regarded digital photography as the enemy, an evil juggernaut that would kill the chemical-based film and paper business that fuelled Kodak's sales and profits for decades" (Mui, 2012). And so, Kodak failed to transform its organization to digital, stating to their own inventor of the first digital camera: "that's cute—but don't tell anyone about it" (Mui, 2012). Kodak's bankruptcy was ultimately the result of its inability to adapt.

Endless and continuous sharing of digital photographic images is now feeding algorithms, enriching big data, and guzzling energy like never before. The consequences of image consumerism are overwhelming: socially, environmentally, politically, and often also mentally. Moreover, with the camera as a tool for visualising socio-political power structures of defining and categorising, photography created a visual standard, established the dominant visual norm. Author, curator and theorist Ariella Aïsha Azoulay (2018, 2020, p. 282) reminds us that "the right to take photographs was imposed from the start as given, unlimited and inalienable, often against the will of others." How to continue my own relationality with photography and at the same time unlearn this "unlimited right to take photographs"? How to find possibilities for systemic change and (self)care among photography's ruins? A way to start is through staying in the process, right in between "You press the button" and "we do the rest". It is an area that textile designer and researcher Svenja Keune (2021) describes as "a way of being-with and staying-with, rather than as a solution-driven practice" or, as theorist and philosopher Donna Haraway (2016) politically situates it: "Staying with the trouble". Here the intersectional figure of radical emergence enters the scene. This figure is not a method or a clearly defined concept, but a way to understand that methods like *unlearning* and *uncertainty* are necessities for practice-based art and design research. A figure, in which 'make-ability' is about sowing seeds.

Humbleness and Wonder

This requires a return to the principle of photographic perception. In the 4th century BC, Chinese philosopher Mo Tzu gave one of the earliest written observations of a camera obscura (Guarnieri, 2016). In a dark room (a camera obscura), Mo Tzu observed the inverted image of an illuminated object outside from which light was reflected into the room through a tiny hole. This phenomenon

369

of reflecting light is where the essence of photography is rooted, in the Greek *phōs*, 'light': light that is reflected when it meets a surface, light that transitions the Silver-nanoparticles in the film emulsion to form an image, light that transfers the negative into a positive. Light as the ultimate source of life on this planet, captured through the process of photosynthesis. Already from photography's technical invention in the early 1820's onward, living materials have been used in photographic processes. Starting with the Anthotype process (Fabbri, 2021), in which light-sensitive emulsions are made from plant-based material, to the Bactograph (Landry, 2012) and Algae-graphies (Giraud, 2011-2014), where bacteria or algae transfer photographic images on agar plates, to Pelargonium Printing (Beaumont-Thomas, 2015; Potter, 1976), where negatives are printed into geranium leaves, to Phytograms (Doing, 2016), where plants directly imprint the film material, to the use of grass as a living photographic medium (Ackroyd & Harvey, 1990). What unites these organic practices is that the human makers determine the visual outcome.

With "commitments to generosity and humility" (Ventin, 2023; Wakkary, 2021) towards Cyanobacteria as our ancient ancestors, I investigate an alternative photographic method where I, as a photographer, am no longer in control of the actual image, but rather facilitate the circumstances and conditions in which photographic images may evolve. It is an attempt to stay in the process, in the messy stuff where a sense of wonder thrives, where the imaginary as a source for 'otherwises' takes over. This type of photography is a "living-with and designing-with" Cyanobacteria, "which refers to a relational and expansive practice of designing in which humans are neither central nor exceptional but rather are ecologically interdependent with the nonhuman world" (Tomico et al., 2023, Wakkary, 2021). In the initial practical application of this 'photographing-with,' endless encounters between *Synechocystis* and photographic film material were observed. As a new approach to analogue photography, *Synechocystis* dissolved the gelatine layer of different black-and-white film emulsions and entered into a dialogue with the film material presented to them.

370

What does it mean to take "response-ability" (Haraway, 2016, p. 28) as a photographic facilitator? "How to address that response-ability (which is always experienced in the company of significant others, in this case, the [Cyanobaceria])?" (Haraway, 2008, p. 89). Do I try to exclude myself from all the actions taken or do I embrace the role as a marked factor of influence in the process? The latter being more realistic and in line with the essence of a collaboration, it was my decision to provide the set-up for the experiments, perform the manoeuvres and then determine the moment when the photographic film material was removed from the Cyanobacteria's habitat. This was my contribution in the role of photographic facilitator in the following experiments.

Case Study #1

In 2021, ignorant but full of courage, I began my first experiment in what I now know is called an 'open habitat'. I slipped an undeveloped black-and-white film strip into a glass tank containing approximately five litres of *Synechocystis* sp. PCC 6803. After ten days, I removed the strip from the Cyanobacteria's habitat and saw that the emulsion had been affected. The gelatine layer of the photographic emulsion seemed to have dissolved into pleated structures. In addition, the *Synechocystis* themselves had also accepted the celluloid as a place to dock, which created a green fog along the spots where the emulsion had coalesced. I thought of the scale ratio of one *Synechocystis* cell and saw a massif of snow-capped mountains looming before me. This negative was our number one, a first outcome. The *Synechocystis* culture fluctuated a bit back and forth around their initial bright green colour, but remained inside the green range, indicating that the cells were alive and dividing.

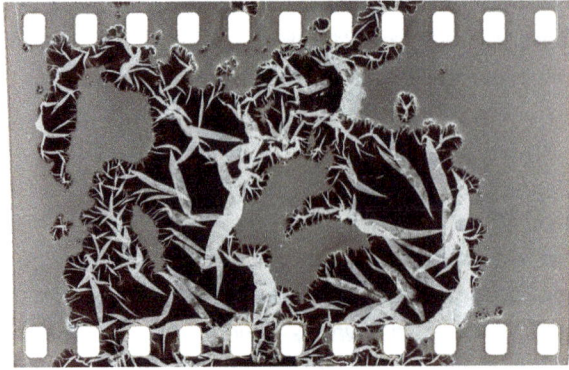

Figure 20.1. Synechocystis sp. PCC 6803, negative number 1.

Figure 20.2. Examples of negatives made by *Synechocystis* sp. PCC 6803.

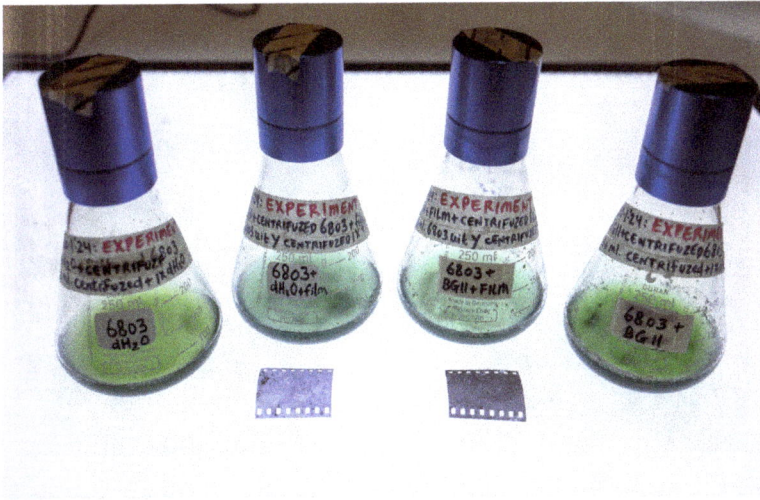

Figure 20.3. Centrifuged *Synechocystis* sp. PCC 6803 in 50ml dH2O or BG-11 medium; with and without B&W film material.

In continuation, various set-ups (and control set-ups) were built and tested, but as of now, many of the bright green Cyanobacteria cultures I introduced the photographic material to turned into watery brownish-yellow substances. The living Cyanobacteria cells transformed into non-living cells: from alive to dead in our limited human understanding, from collaboration to material. The switch from green to yellowish-brown continues to move me. It makes me realise that part of working together with an organism is equality, that as a human being, if I consider myself akin to Cyanobacteria, I must confront the fact that this loss of cells is a direct consequence of my actions. However, the insight that I can mourn the loss of a culture of *Synechocystis*, is the consequence of the many mistakes I made, it is the realisation of the failed experiments and not so much of the successful ones.

Figure 20.4. Living and non-living *Synechocystis* sp. PCC 6803 cultures at STROOM, The Hague, Netherlands.

Case Study #2

After the experiments within *Synechocystis*'s liquid habitat, we moved to the semi-solid surface of the agar plate. Time and again, agar plates were prepared and inoculated in the Material Incubator Lab in Den Bosch and then moved to my studio in The Hague. A variety of possible 'othernesses' were tested, for example: agar with and without BG-11 medium; with and without photographic film material; emulsion up or down; film with and without emulsion; and different positions of the film and of *Synechocystis* (in, under and on the agar or film). The Cyanobacteria were even centrifuged with sterile dH2O to get the BG-11 medium they lived in as much as possible out of their system before inoculation. When the agar plates arrived in my studio, they were provided with as optimal growing conditions as possible considering the circumstances. Some general observations are that the process of diluting the photographic emulsion took significantly longer than in the liquid medium and that the place of inoculation determined the traces *Synechocystis* left on the photographic material. More specific

374

observations are that without BG-11 medium, the emulsion of the film material is diluted more quickly, and that *Synechocystis* grows more concentrated on the intersection of agar and film material (with or without the emulsion layer).

During the period of several years that these experiments were performed, it also became clear that the final result is not what is at stake. Obtaining an aesthetically interesting negative stands in no relation to the loss of a culture of Cyanobacteria. Reflecting on our collaboration, it has become clear that it is not so much about utilising Cyanobacteria to create photographic images, but much more about keeping *Synechocystis* alive through the deconstruction of photographic material. Jointly we work on the decomposing of the photographic film emulsion by engaging in a form of bioremediation. This reversal of interest emphasises the urgency of staying in the process versus the pursuit of a possible outcome.

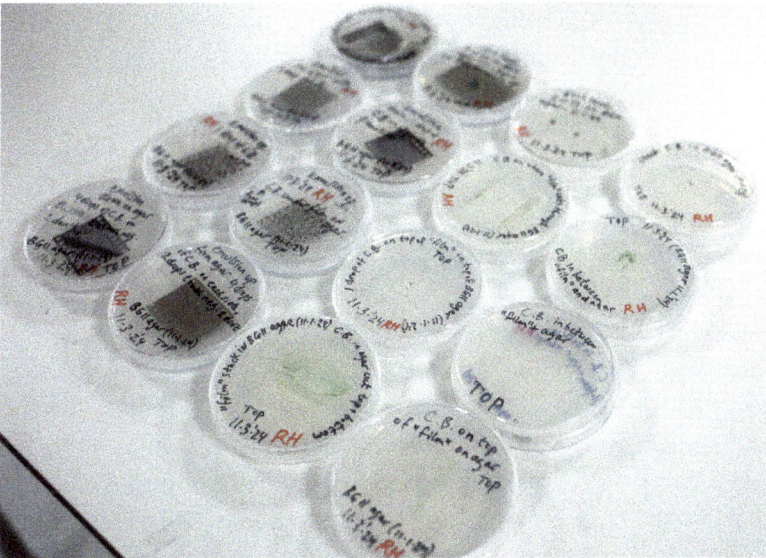

Figure 20.5. Agar plates with film (with and without emulsion) inoculated with Synechocystis sp. PCC 6803 in Material Incubator Lab, Den Bosch, Netherlands.

Reflection

These experimental exercises in decentralizing human experience to imagine possible mutualistic alliances between both the human and the non-human, both the living and the non-living, challenge the understanding of what a photograph is. Here I build upon the notion of "mutualistic care", one of the three "fundamental loci of designing for livingness", "where *livingness* is understood as a biological, ecological, and experiential phenomenon" (Karana et al., 2020). And "where mutualistic care represents (…) various forms of reciprocity that occur during interactions among multiple species" (Groutars et al., 2024). As a next step to move away from the photographic event with an image as only possible result and towards a living photograph, Cyanobacteria will no longer be the responder to photographic film material but rather be the light-sensitive medium itself. As ubiquitous organisms, Cyanobacteria are like a perpetual living photographic process, performing both the process of photosynthetic sensing of the sun and the process of cell division: a continuum of becoming, a figure of radical emergence.

I needed a few years to really understand how futile I am in the life—or for that matter the non-life—of Cyanobacteria. A fleeting passer-by, an awkward interrupter who wants their actions to affect the organism they wish to collaborate with. Now I wonder: what would happen if agency were to be truly ceded to the more-than-human? What are bacteria, algae, grass, or Cyanobacteria teaching us? (Kimmerer, 2013). What kind of 'make-ability' are they applying on a daily, yearly, millennially, eonly basis? Are we able to go beyond our human understandings when we interpret the processes created by other(s)-than-human? Are our eyes, or, for that matter, all our senses together, sensitive enough to grasp the complexity of what we encounter?

Key Takeaways

Practice-based art and design research are constantly in flux as an intersectional figure of radical emergence. Changing research attitudes from make-ability to response-ability, help facilitate a movement towards ceding agency to the more-than-human.

- Human action always moves within the context of more-than-human prosperity.
- Facing toxic histories implies feeling addressed.
- "How do we do what we do humble, accountable and in good relation?" (Liboiron, 2021b).

References

- Aiyer, K. (2022, February 18). The Great Oxidation Event: How cyanobacteria changed life. *American Society for Microbiology*. https://asm.org/articles/2022/february/the-great-oxidation-event-how-cyanobacteria-change

- Anthony, S. D. (2016, July 15). Kodak's downfall wasn't about technology. *Harvard Business Review*.https://hbr.org/2016/07/kodaks-downfall-wasnt-about-technology

- Azoulay, A. A. (2018, October 10). Unlearning imperial rights to take (photographs). *Verso Books*.https://www.versobooks.com/blogs/news/4075-unlearning-imperial-rights-to-take-photographs

- Azoulay, A. A. (2020). *Potential history: Unlearning imperialism.* Verso.

- Beaumont-Thomas, B. (2015, September 24). Alice Cazenave's best photograph: A portrait on a leaf. *The Guardian.* https://www.theguardian.com/artanddesign/2015/sep/24/alice-cazenave-best-photograph-portrait-on-leaf-photosynthesis

- Borgmann, A. (1984). *Technology and the character of contemporary life.* University of Chicago Press.

- Emerson, R. W. (1841). *Self-reliance.* https://archive.vcu.edu/english/engweb/transcendentalism/authors/emerson/essays/selfreliance.html

- Fabbri, M. (2021, April 7). The history of anthotypes. *Alternative Photography.*https://www.alternativephotography.com/the-history-of-anthotypes/

- Fineman, M. (2004, October). Kodak and the rise of amateur photography. *The Metropolitan Museum of Art.*https://www.metmuseum.org/toah/hd/kodk/hd_kodk.htm

- Gaver, W. W., Gall Krogh, P., Boucher, A., & Chatting, D. (2022). Emergence as a feature of practice-based design research. *Designing Interactive Systems Conference*, June 13–17, 2022, Virtual Event, Australia. ACM. https://doi.org/10.1145/3532106.3533524

- Giraud, L. (2011–2014). *CULTURES - Installation algægraphique.* http://www.liagiraud.com/cultures/

- Groutars, E. G., Kim, R., & Karana, E. (2024). Designing living artefacts for multispecies interactions: An ecological approach. *International Journal of Design, 18*(2), 59–78. https://doi.org/10.57698/v18i2.04

- Guarnieri, M. (2016). The rise of light – Discovering its secrets. *Proceedings of the IEEE, 104*(2), 258–274. https://doi.org/10.1109/JPROC.2015.2513118

- Haraway, D. J. (2008). *When species meet.* University of Minnesota Press.

- Haraway, D. J. (2016). *Staying with the trouble: Making kin in the Chthulucene.* Duke University Press.

- Hazekamp, R. (2023). Cyanobacteria stories: Moving inside out of Elizabeth Povinelli's 'Carbon Imaginary.' *Trigger#5, FOMU, Antwerp*, 14–23.

- Karana, E., Barati, B., & Giaccardi, E. (2020). Living artefacts: Conceptualizing livingness as a material quality in everyday artefacts. *International Journal of Design, 14*(3), 57–73.

- Keune, S. (2021). Designing and living with organisms: Weaving entangled worlds as doing multispecies philosophy. *Journal of Textile Design Research and Practice, 9*(1), 9–30.

- Kimmerer, R. W. (2013). *Braiding sweetgrass: Indigenous wisdom, scientific knowledge, and the teachings of plants.* Milkweed Editions.

- Liboiron, M. (2021a). *Pollution is colonialism.* Duke University Press.

- Liboiron, M. (2021b, November 3). Why pollution is as much about colonialism as chemicals. *Don't Call Us Resilient Podcast*, S2:E5. https://dont-call-me-resilient.simplecast.com/episodes/ep-11-why-pollution-is-as-much-about-colonialism-as-chemicals

- Lloyd, K. (2022, July 31). Undead matter #3: Existing between | Radio Web MACBA | RWM podcasts. *MACBA Museu d'Art Contemporani de Barcelona.* https://rwm.macba.cat/en/podcasts/undead-matter-3-existing-between/

- Mui, C. (2012, January 18). How Kodak failed. *Forbes.* https://www.forbes.com/sites/chunkamui/2012/01/18/how-kodak-failed/

- Potter, G. (1976). The natural history of a sunbeam – A leaf from nature. *The Royal Institution.*https://www.rigb.org/explore-science/explore/video/natural-history-sunbeam-leaf-nature-1976

- Povinelli, E. A. (2016a). *Geontologies: A requiem to late liberalism.* Duke University Press.

- Povinelli, E. A. (2016b). Fires, fogs, winds. In *Elements for a world: Fire* (pp. 21–29). Published in conjunction with the exhibition *Let's Talk About the Weather: Art and Ecology in a Time of Crisis*, Sursock Museum, Beirut, Lebanon.

- Povinelli, E. A. (2021). *Between Gaia and ground: Four axioms of existence and the ancestral catastrophe of late liberalism.* Duke University Press.

- Puig de la Bellacasa, M. (2017). *Matters of care: Speculative ethics in more-than-human worlds.* University of Minnesota Press.

- Robbins, H., Giaccardi, E., & Karana, E. (2016). Traces as an approach to design for focal things and practices. *NordiCHI '16: Proceedings of the 9th Nordic Conference on Human-Computer Interaction*, Article No. 19, 1–10. https://doi.org/10.1145/2971485.2971538

- Tomico, O., Wakkary, R., & Andersen, K. (2023). Living-with and designing-with plants. *Interactions, 30*(1), 30–34. https://doi.org/10.1145/3571589

- Ventin, M. (2023, September 27). Ron Wakkary. *Design-Decode.* https://www.designdecode.org/ron-wakkary/

- Wakkary, R. (2021). *Things we could design: For more-than-human-centered worlds.* MIT Press.

- Zhou, J., Doubrovski, Z., Giaccardi, E., & Karana, E. (2024). Living with cyanobacteria: Exploring materiality in caring for microbes in everyday life. *CHI '24: Proceedings of the 2024 CHI Conference on Human Factors in Computing Systems*, Article 561, 1–20. https://doi.org/10.1145/3613904.3642039

Koen van Turnhout, Peter Joore,
Remko van der Lugt, Troy Nachtigall,
Liliya Terzieva

21. Epilogue: Embracing Design as a Collective Learning Journey

DOI: 10.1201/9781003609766

Embracing Design as a Collective Learning Journey

Applied design research thrives in and even flourishes in complexity. It is more than just a methodology, a lens or a problem-solving approach—it is a transformative practice that shapes, nourishes and intertwines technology, societal structures, and professional knowledge ecosystems. Rather than avoiding complexity, applied design researchers embrace it as a space and continuum for inquiry and innovation. As Kuipers et al. (Chapter 4) describe, applied design researchers create a temporary boundary world. A fluid space where boundaries shift, and participants discover new ways of seeing, understanding, and responding to the changes we aspire to achieve.

In this book, multiple authors present perspectives on how applied design research projects intersect with the rapid changes occurring around us. What is being verified and explored through the diverse chapters boils down to diverse insights. Technologies and practices, as seen, do not emerge in isolation; they evolve along trajectories shaped by social, economic, and cultural forces. Applied design research is fundamentally about facilitating transitions within these tension fields. They sensitise to societal changes as they unfold and intervene at precisely chosen leverage points. This requires identifying the most pressing questions and uncovering the points of highest impact—where insights have the power to shift perspectives and drive meaningful change.

A systemic perspective can be valuable in this context. Several chapters in this book explore how designers can navigate the complexities of interconnected systems involving people, environments, and technologies. Applied design researchers often initiate change through small-scale or micro-level interventions aimed at driving broader systemic transitions. While these interventions can be effective when well-chosen, many chapters suggest they are just the starting point. Lasting change requires engagement at multiple system levels—something that cannot be achieved in isolation: success depends on collaborating with stakeholders who may initially view design and design research as unfamiliar or even foreign concepts.

As such, applied design research projects must serve as safe spaces for stakeholders from diverse backgrounds and value systems. In a way, these projects act as microcosms of the broader societal change they seek to inspire—using design to reshape ways of thinking. This means applied design research projects need to accommodate a wide range of stakeholders, with all kinds of backgrounds and value systems. Rather than positioning designers as solitary creators, applied design research frames them as facilitators within learning collectives. The ambition is to cultivate environments that foster learning, critical reflection, and societal transformation. Many chapters in this volume highlight the importance of creating inclusive environments where everyone has the ability to change everyone.

Taking this narrative of collective learning seriously, we also need to acknowledge the dynamically changing role of design in society. The impact of applied design research lies not in producing isolated solutions to obvious problems but in navigating the psychological and societal dimensions of change. Stakeholders, non-designers, are invited to see the world through new lenses, expand their solution repertoires, and to challenge their assumptions—in particular about the future. More than simply solving discrete problems, applied design researchers engage societal players in transformative journeys. Any solutions that emerge in applied design research projects are merely waypoints, not final destinations. It is not about the specific route taken, but about the broader path of exploration and discovery that unfolds. Design, so to say, has no end.

What are, after saying all of this, open avenues for future research? We suggest a couple of directions.

- What are the 'sweet spots' of intervention with design research? Should we be looking at how society evolves and use that as an anchor point, should we address the most pressing issues or should we aim for iconic results, and the impact they have on collective narratives?
- Is shifting system levels an opportunity or a cost? It is clear that design can navigate multiple system layers and that it can boil down abstract difficulties into concrete solutions at a lower level of abstraction. But how does it feed back into the larger system, and how can we be empowered to support the movement towards 'higher levels' of 'the system'?
- How do we scale the lessons learned from breaking and changing disciplinary boundaries within a project to society as a whole? Can we change temporary boundary worlds into permanent ones?
- How do we navigate tangibility and fluidity, which are both considered characteristics of applied design research? Is the linguistic, practical or ideological vocabulary that we develop within a project a prototype for something bigger in society or does it remain within the boundaries of the project?
- If 'everything is fluid', apparently we may conclude that the goal of a project is mostly about the broader path taken, and much less about the concrete results of a project. How do we boil that down to concrete projects with finite deliverables?
- If we are so divergent about what counts as 'success' or 'impact' in an applied design research project, how do we make sense of, define and defend the impact of our projects? Do we refer to a multidimensional impact measurement, or are there ways to avoid the perils of impact measurements?

Koen van Turnhout, Peter Joore, Remko van der Lugt, Troy Nachtigall, Liliya Terzieva

About the Editors

386 DOI: 10.1201/9781003609766

Koen van Turnhout

Koen van Turnhout is professor of Human Experience & Media Design at Utrecht University of Applied Sciences. His research group examines the design of digital media. In particular, how the solution repertoire of digital designers can be strengthened for new technological developments such as AI and immersive media. A special research focus is the ubiquity of media and how designers can design for sensitive and emotional use-situations such as mental health or political participation. Koen has a background in human-computer interaction, multi-modal interaction, mixed realities, design, and research methodology. He is editor of several books on design research, including *Applied Design Research in Living Labs and Other Experimental Learning and Innovation Environments* and the *Handboek Ontwerpgericht Wetenschappelijk Onderzoek.*

Peter Joore

Peter Joore is professor of applied sciences and chair of the Open Innovation Research group at NHL Stenden University of Applied Sciences, Leeuwarden, The Netherlands. He is chair of the Network Applied Design Research and holds a PhD from the faculty of Industrial Design Engineering of Delft University of Technology, where he is still a visiting researcher at the Design for Sustainability section. In all his work, his main aim is the realization of a smart, sustainable, inclusive society. Peter's research focuses on the relationship between long-term societal transition processes on the one hand, and short-term product and service innovations on the other.

Remko van der Lugt

Remko is Professor of Co-Design at Utrecht University of Applied Sciences. His research focuses on involving people in collaborative creative design activities, enabling them to participate as experts of their experiences. He has a particular interest in systemic design, particularly supporting the ways in which designers can facilitate change processes by means of extending their repertoire of tools, methods and skills. In this he connects the domains of participatory design with systemic work, with tools like socionas mission mapping and design probing. Remko has broad experience in facilitating creative processes in organisations, both in business and government.

Troy Nachtigall

Troy Nachtigall is an experienced Designer and Design Researcher who leads research at the intersection of fashion and technology. He actively seeks out strategic partnerships that will play a crucial role in shaping the future of the textile, clothing, fashion, and footwear industries. Troy is a core member of the Erasmus+ Fashion and Textile Transitions project and the New Textile Ecosystems (NewTexEco) Research Community. His practical background in Design and ICT fuels his passion for exploring the data-material relationship through wearable items. As a Professor at Amsterdam University of Applied Sciences, he heads the Wearable Data Studio, which focuses on the computational fabrication of clothing, shoes, and accessories. Through this research, Troy and his team can delve into the complex adaptive ecosystem of fashion, focusing on data management, design, manufacturing, and use. This has led to the development of physical, digital, and hybrid artifacts and toolkits that promote a material perspective and data-enabled ecosystem lifecycle frameworks.

Liliya Terzieva

Liliya Terzieva is Professor of Designing Value Networks at the Hague University of Applied Sciences, the Netherlands. Her PhD is in the field of Economic and Organizational sciences of Leisure and Tourism. Within her career pathway of around 20 years of international experience (Bulgaria, China, France, Malta, Vietnam, the Netherlands, etc.) she worked in non-governmental, public, as well as educational and business organizations, related to the domains of (among others): entrepreneurial learning and mindset, value generation, leadership, tourism, leisure, strategic and human-centred design. Her research interest is in the impact and value that diverse multi-stakeholder collaborative interactions lead to, and the role design plays in fostering and nurturing the above.

About the Authors

390 DOI: 10.1201/9781003609766
This chapter has been made available under a CC BY-NC-ND license.

Daan Andriessen

Daan Andriessen is professor of Research Competence at University of Applied Sciences Utrecht. He specializes in optimizing the impact of practice-based research and the competence development of researchers at universities of applied sciences. He has a background in administrative science, methodology and knowledge management and a PhD in Economics from Nyenrode University. His mission is to support researchers in doing research that matters; practice-based research that contributes to the development of knowledge, people, products, and organisations. He has a special interest in the dilemma between rigour and relevance.

Danielle Arets

Danielle Arets is a professor of Design in Journalism at Fontys University of Applied Sciences in Tilburg. The professorship focuses on journalistic transformation with an emphasis on digital developments, new forms of public engagement, and the (re)design of journalistic storytelling and expression. Before joining Fontys, Danielle was an associate professor of Strategic Creativity for 8 years at the Design Academy Eindhoven. In addition to her work at Fontys, Danielle is a member of the Dutch Council for Culture, a member of the advisory board for Click NL, PONT and NEWS.

Bas de Boer

Bas de Boer is a philosopher of technoscience who works as an assistant professor at the University of Twente, The Netherlands. Currently, his research focuses on how technologies shape the experience and understanding of health. His further research interests are in phenomenology, philosophy of technology, and philosophy of health and medicine. He is the author of *How Scientific Instruments Speak: Postphenomenology and Technological Mediations in Neuroscientific Practice* (Lexington Books) and *Healthy Embodiment: Philosophical Reflections on the Experience of Health* (Routledge). His research has appeared journals such as *Design Issues, Medicine, Healthcare and Philosophy, and Phenomenology and the Cognitive Sciences.*

Aranka Dijkstra

Aranka Dijkstra looks to accelerate sustainable transitions through implementing real-life experiments and interdisciplinary collaborations. Since 2014, she has been guiding students and entrepreneurs in designing and implementing experiments in various urban and festival living labs. Even after many years, Aranka maintains a strong connection to festival experimentation and aims to spread her knowledge and experience with festival experimentation through various publications and in particular the Festival Experimentation Guide, a practical guide on how to experiment at festivals created in collaboration with Marije Boonstra, Peter Joore and with the support of NHL Stenden University of Applied Sciences.

Agnes Evangelista

Agnes Evangelista is a Junior Researcher with the Circular Business Research Group, focusing on sustainability and circular economy models. She holds a Bachelor's degree in International Public Management from The Hague University of Applied Sciences and a History degree from Andrew College. With a strong background and passion for academic writing and research, she has co-authored journal articles and contributed to international conferences. Her work supports business model research, sustainable fashion, ways to apply generative AI in co-crafting sustainable futures as well as interdisciplinary approaches to sustainability.

Poorvi Garag

Poorvi Garag is a strategic social designer and design researcher working on projects dominantly focused on community, policy and resources in the public sector. With a deep commitment to creating social impact and change, she works with industrial design and design research skills informed by anthropology, sociology, futurology, and decolonial studies. She graduated with a Social Design Masters at Design Academy Eindhoven and an Industrial Design Bachelor at Srishti Institute, Bangalore.

Christa van Gessel

Christa van Gessel is a design researcher at the Co-Design research group at HU University of Applied Sciences Utrecht. As a co-designer and researcher, Christa brings people together in complex projects to work in a designing way and to reflect on the method together. Her research focus is on generative tools to enable people to co-research and to co-design their own context together. She studies how people, not educated as designers, learn, apply, and feel confident with design tools, and how these generative tools can carry and make the knowledge usable within the project. Christa studied Industrial Design Engineering and Strategic Product Design at the Delft University of Technology and is experienced in designing, researching, consulting, and teaching.

Jetse Goris

Jetse Goris (1980) Jetse Goris is an educational consultant and innovator, focusing on human-centered design in healthcare and medical education. He was part of the team that formed the Innovation Support Center of the University Medical Center of Groningen and lectured on design-driven innovation at the University of Applied Sciences NHL Stenden. Currently he is a freelance design-driven innovation advisor and freelance (talk show) host at academic events.

Kees Greven

Kees Greven is a researcher at the Utrecht university of applied sciences (HU). He has a background in philosophy of science, and his research topics at the HU include ethics and philosophy of science of AI, methodology of practice-based research, complexity theory, art-based research, and impact evaluation. Additionally, he holds workshops and lectures at the HU about philosophy of science and ethics in the context of practice-based research.

Risk Hazekamp

Risk Hazekamp (they/them) is an inter-dependent visual artist, researcher and art educator. After studies at Willem de Kooning Academy (Rotterdam) and Jan van Eyck Academy (Maastricht), they lived in Berlin for 11 years. Hazekamp graduated in 2020 at the 'Advanced Master of Research in Art & Design' at St Lucas School of Arts Antwerp. In 2023, Hazekamp started as a Professional Doctorate candidate at the 'Centre of Applied Research for Art, Design and Technology' (CARADT/Avans) in Den Bosch, in collaboration with the 'Materials Experience Lab' at TU Delft and the Professorship 'Art and Sustainability' (KCKS/Hanze) in Groningen.

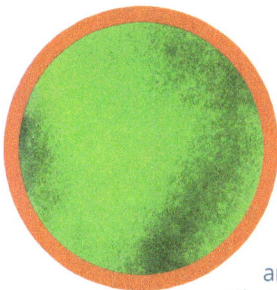

Synechocystis sp. PCC 6803

Synechocystis sp. PCC 6803 is a strain of freshwater Cyanobacteria. Cyanobacteria were the first organisms on planet Earth to perform oxygenic photosynthesis. Synechocystis are single-celled prokaryotes, which means that they do not have a distinct nucleus with a membrane. The original strain was first isolated in 1968 by Riyo Kunisawa from a freshwater lake near Berkeley, California. In a human-centered context, Synechocystis can be grown on either agar plates or in liquid culture. Under the microscope, humans are able to see this unicellular ancient ancestor as a little round green dot.

Jan-Wessel Hovingh

Jan Wessel Hovingh (1977) is a graphic, interaction and UX designer by trade and has amassed over 25 years of experience in the field. At present, he teaches design and design research at NHL Stenden University in Leeuwarden, the Netherlands. Concurrently, he is pursuing a doctoral degree at the University Medical Centre in Groningen, also in the Netherlands, with a research focus on the moral and ethical dimensions of technological mediation and the societal role of digital products in human society.

Tomasz Jaskiewicz

Tomasz Jaskiewicz is Professor (lector) of Civic Prototyping at the Rotterdam University of Applied Sciences and Design Fellow in Prototyping Complexity at the Faculty of Industrial Design Engineering at the TU Delft. Tomasz' core belief is that technological innovation should be an inclusive and democratic process. To this end, his research revolves around developing new tools, methods and strategies supporting designers and non-professionals in iterative prototyping with digital technologies. His work is strongly influenced by his background in architecture and urban design, experience of running and failing a startup, and a life-long passion for building quick-and-dirty interactive prototypes.

Derek Kuipers

Derek Kuipers (1975) is Professor of Design Driven Innovation at NHL Stenden University in Leeuwarden (NL). Kuipers has a background in educational sciences, with over 25 years of experience. His research has focused on digital media and technology, with a particular emphasis on design processes and serious gaming. This research led to his PhD thesis, 'Design for Transfer', published in 2019. Professor Kuipers is a distinguished member of the academic community, serving as a founding faculty member of three master's degree programs at NHL Stenden University: Health Innovation, Serious Gaming, and Design Driven Innovation.

Elise van der Laan

Dr. Elise van der Laan is a sociologist and Associate Lecturer in Fashion & Design Research at ArtEZ University of the Arts. She obtained her PhD (2015) from the University of Amsterdam for a study into aesthetic standards in fashion photography. She has published on topics such as fashion and identity, aesthetic standards, sustainable fashion, cultural fields and living labs. She currently specializes in applied research around regenerative fashion and care and repair of clothing.

396

Mailin Lemke

Mailin Lemke worked as a design researcher at the Technical University of Delft in the Netherlands, focusing on the intersection of design and behaviour change interventions. Her work encompassed analysing how designs can influence user behaviour and creating solutions for different user groups that allow and facilitate behaviour change. She now works as a designer and senior design researcher for the design agency dreiform in Cologne.

Marjolein Mesman

Marjolein Mesman (Amsterdam 1966) is fashion retail expert, trainer and part-time lecturer @TMO fashion business school. She teaches Corporate Responsibility and leads the research programs for circular and sustainable business. She has over 25 years of experience in business leadership and sales programs in both wholesale and retail. In her work as a retail trainer, she combines her knowledge of retail with behavioral and communication studies to help retailers optimize their customer journey and improve customer engagement. She believes that the 'art of selling' can be a strong tool in the transition to more sustainable fashion buying behavior; 'creating demand has been the key to successful retail since forever. So why not use it to create good demand'?

Catelijne van Middelkoop

Catelijne van Middelkoop is Head (a.i.) of ArtScience, the Interfaculty of the Royal Conservatoire and Royal Academy of Art in The Hague. Prior to this, she served as Practor (Research Professor) at SintLucas (2020–2024), where she explored 'making' as both a didactic principle and an alternative mode of knowledge creation. In addition, Catelijne is an Design Fellow and Co-Director of the Pictorial Research Lab at TU Delft/IDE. As Co-Director of Strange Attractors, a design research studio founded at Cranbrook Academy of Art (2001), she specializes in developing innovative methods, tools, and critical perspectives that drive measurable change.

Lenny van Onselen

Lenny van Onselen is a design researcher at the Co-Design research group at HU University of Applied Sciences Utrecht. She studies meaningful systemic co-design and developing design ability. She also designs toolkits for professionals and supports the design of hybrid learning environments. Additionally, she assists curriculum developers with the design of hybrid learning environments. Lenny studied Industrial Design Engineering and Strategic Product Design at the Delft University of Technology. She obtained her PhD with a dissertation, *Becoming a design professional through coping with value-based conflicts*. She has experience in consulting, teaching, and researching sustainable co-design and curriculum innovation.

Mieke Oostra

Mieke Oostra is Professor of Applied Energy Transition at the University of Applied Sciences Utrecht. As a lead of the theme Circular and Energy Neutral regions of the Centre of Expertise Smart Sustainable Cities, she focusses on how to future-proof buildings, districts and supply chains in relation to the energy transition of the built environment. Hereto she is creating different transdisciplinary environments to entice researchers, practitioners and citizens to cross boundaries between disciplines, sectors and different knowledge types. Prior to this, she was Professor of Spatial Transformation at Hanze University of Applied Sciences (2012–2018) and Professor of Innovative Technology in Construction at Saxion (2013–2016).

Anja Overdiek

Anja Overdiek currently serves as professor of Cybersocial Design at Rotterdam University of Applied Sciences and as associate professor of Designing Value Networks at The Hague University of Applied Sciences in The Netherlands. Her research focusses on multi-stakeholder Co-Design in processes of societal innovation and transition, on experimental environments, and on digital design for sustainability transitions in the city. Anja co-authored several recent books about Living Labs (LL) and Applied Design Research (ADR) and she is a founding partner of the Expert Network Systemic Co-design (ESC). She holds a PhD from Freie Universität Berlin (Germany) in Political Sciences and approaches her research with a systemic and "more-than-human" design lens.

Kim Poldner

Kim Poldner serves as Endowed Professor of Circular Regional and Economic Development at the University of Groningen in the Netherlands and holds a PhD from the University of St. Gallen, Switzerland. In 2021 she published the Routledge book, *Circular Economy; Challenges and Opportunities for Ethical and Sustainable Business*. Her research interests evolve at the crossroads of entrepreneurship, aesthetics and sustainability and she has written award-winning case studies on sustainable fashion pioneers such as Veja and Osklen. Before she embarked on an academic career, she was a serial entrepreneur in sustainable fashion in the Netherlands, Africa and Brazil.

Margo Rooijackers

Margo Rooijackers is a lecturer and researcher at the Academy for Leisure & Events at Breda University of Applied Sciences (BUas). With a background in psychology, she specializes in consumer behavior, visitor experiences, and creative placemaking. Margo has published research on various leisure-related topics and contributed to a book on *Events as Strategic Marketing Tool*. Additionally, she serves on the editorial committee of the Dutch annual report, *Trendrapport Toerisme, Recreatie en Vrije Tijd*, which covers trends in the leisure and tourism industries.

Angelique Ruiter

Angelique Ruiter works as an associate professor at the Co-Design research group at Utrecht University of Applied Sciences. Their focus is on systems change through the lens of awareness-based systems change and using qualitative and creative methodologies. They are also a teacher, learning team supervisor and curriculum developer at the Master Community Development and the minor Religion, Philosophy and Spirituality. Angelique studied European Studies at the University of Amsterdam and obtained their PhD in sociology at Humboldt Universitat zu Berlin. Their dissertation focused on cultural hybridity in relation to subcultures in Berlin.

Shakila Shayan

Shakila Shayan is a senior researcher in Human Experience and Media Design and a lecturer for the Master in Data Driven Design at Hogeschool Utrecht (HU). In her research she focuses on application of conversational AI in healthcare. She holds a PhD in Cognitive Science and Computer Science from Indiana University in the United States, and has conducted research at Max Planck Institute and Utrecht University. Before joining HU, she completed a fellowship in Social Innovation in London, working with the education sector and social housing associations. Originally from Iran, she earned her B.Sc. in Computer and Electrical Engineering from the University of Tehran.

Wina Smeenk

Wina Smeenk is professor of Societal Impact Design at Inholland University of Applied Sciences. She is founder and chair of the Dutch Expertisenetwork Systemic Co-design (ESC). Wina graduated from Delft University of Technology, where she studied Industrial Design, after which she spent over 25 years working as a co-designer and design researcher for international businesses, government and non-profit organisations. In 2010, Wina launched the empathic co-design agency 'Wiens Ontwerperschap'. She co-developed many design-oriented educational programs. In 2019, she defended her PhD thesis 'Navigating Empathy, empathic formation in co-design processes'. Apart from academic articles such as the Empathy Compass, Wina co-edited the book *Applied Design Research in Living Labs and Other Experimental Learning and Innovation Environments* and authored the books *The Co-Design Canvas* and *Design Play Change*.

Iskander Smit

Iskander is the founder and chair of the Cities of Things Foundation, a research program that started at TU Delft, where he was a visiting professor. It now serves as an independent knowledge platform in cooperation with academic partners and industry. Iskander is deeply involved in exploring the relationship between human and machine intelligence from a design perspective. He founded Target_is_New, a futures research practice and weekly newsletter on human-AI interplay, and is a curator and board member of ThingsCon. Before this, he held positions in strategy and innovation and was involved in various technology initiatives and events as an organizer and public speaker.

400

Guido Stompff

Guido Stompff is a professor of Applied Sciences at Inholland University of Applied Sciences. He specializes in design thinking/doing and advocates for the relevance of design skills for any professional. His research focuses on finding pathways of action to democratize design. In his design career, spanning 25 years, he won numerous design awards. In his academic career, he obtained a PhD on innovation-in-the-wild, with a focus on collective intelligence. He leads a team that focuses on participatory approaches to innovation. He is engaged in networks as Expertisenetwerk Systemisch Codesign (ESC) and the Center of Creative Innovation (CoECI)..

Gijs Terlouw

Gijs Terlouw is an Associate Professor of Applied Sciences in Health Innovation and Simulation Learning at NHL Stenden. His research is characterized by a design-driven approach, focused on improving healthcare and healthcare education. Within the context of healthcare, boundary objects, simulations, and serious games are recurring themes throughout his research career.

Jet van der Touw

Jet van der Touw is a researcher in Human Experience and Media Design at the University of Applied Sciences Utrecht. In her research, she focuses on co-design methods and how these can be applied for impactful and empathic design solutions. She earned her master's degree in Design Research, where she researched the topic of home and belonging, influenced by her double nationality of Croatian and Dutch roots. Prior to that she completed a Bachelor of Art & Design at the Gerrit Rietveld Academy.

Lars Veldmeijer

Lars Veldmeijer is a PhD candidate at University Medical Center Utrecht, and a lecturer and researcher at NHL Stenden University of Applied Sciences. In his PhD project, he combines his lived experience of mental distress with his background in design for health. His main focus is exploring the potential of design activity and co-creation in mental health care.

Sophie Vermaning

Sophie Vermaning is a lecturer-researcher at the Creative Media for Social Change (CMfSC) research group at the Amsterdam University of Applied Sciences (HvA) and a team leader in the Communication and Multimedia Design (CMD) program. With a background in Pedagogy, she focuses on social issues and promoting inclusivity. In her work, she combines media, education, and social impact to develop innovative solutions. Through her roles in both education and research, Sophie contributes to the development of students and professionals who use media and technology for social progress. Her passion lies in connecting creativity with social justice.

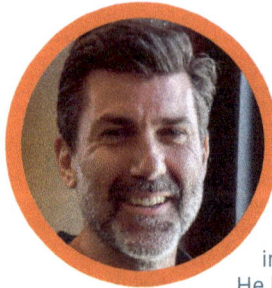

Nick Verouden

Nick Verouden is an associate professor of Creative Media for Social Change at the Amsterdam University of Applied Sciences. With a background in anthropology, he applies design ethnographic and artistic approaches to examine and shape participatory processes, engaging diverse audiences in debates on climate change and sustainability. He holds a PhD from Delft University of Technology, where he studied conversations within complex collaborative processes, focusing on the significance of issues that remain unspoken, overlooked, or silenced.

Rens van der Vorst

Rens van der Vorst is a Dutch techno-philosopher, educator, and author based in Breda. He serves as a lecturer and researcher at Fontys University of Applied Sciences, guiding ICT students in examining technology's impact on society. Renowned for his engaging presentations that blend humor with critical insights, Rens encourages audiences to reflect on their relationship with digital technology. He has authored several books, which explore the pervasive influence of technology on daily life. Additionally, he developed the Technology Impact Cycle Tool, a globally utilized resource designed to assess the implications of technological innovations.

Inge Vos

Inge Vos is Program Manager at the Stichting het Gehandicapte Kind, where she is responsible for overseeing various projects under the umbrella of the Samen Sporten program, which fosters interactions between children with and without special needs by facilitating sports activities at local clubs. Previously, she spent over 17 years at patient association Spierziekten Nederland, developing innovative care programs and resources for children with neuromuscular diseases and their families. Inge was also a Researcher at the Vrije Universiteit Amsterdam within the Department of Metamedica and Medical History. Inge holds a master's degree in Special Education/Orthopedagogy.

Jeroen de Vos

Jeroen de Vos is a media anthropologist and researcher at the Designing Journalism research group at Fontys University of Applied Sciences for Journalism. With a unique combination of qualitative research, technology philosophy and social media data analysis, he brings together human and technological perspectives. Jeroen prefers to work in the interdisciplinary grey areas where innovation and creativity flourish. He structures chaos, finds common ground and makes complex issues manageable. In addition to being a researcher at Fontys, he is also a research associate at the Faculty of Media Studies at the University of Amsterdam.

Roelof de Vries

Dr. Roelof Anne Jelle de Vries is a senior researcher and lecturer at Utrecht University of Applied Sciences. Within the Human Experience & Media Design research group, he investigates the interdisciplinary design process of interactive technology and its evaluation and validation. How can we design interactive technology in such a way that it helps with the growing complexity of health and welfare issues? He is originally a computer scientist and did his doctoral research on theory-based and personalised behavioural change technology. He has experience in design-oriented scientific research, in both research and design methodologies.

Judith Weda

Judith Weda has been working as a researcher in the Human Experience & Media Design and Smart Systems for Healthy Living research groups at Utrecht University of Applied Sciences since September 2024. She holds a PhD in Human Computer Interaction and her thesis explored the experience and perception of pressure on the skin caused by soft actuators. She conducted this research within the WEAFING consortium, which aimed to develop active textiles for wearable technology that would allow you to give your grandmother a hug during a video call, for example. Her research interests include human-computer interaction, wearables and haptics.

Bart Wernaart

Bart Wernaart (LL.M., PhD) is professor of Moral Design Strategy and chair of the Center of Expertise for Sustainable and Circular Transitions for Fontys University of Applied Sciences, the Netherlands. He is an author in the field of international human rights law, moral design and (business) ethics. He is also a member of the ethics committee for the municipality of Eindhoven. Next to his academic career, he is a professional drummer, conductor and composer.

Tamara Witschge

Tamara Witschge (PhD University of Amsterdam) is professor of Creative Media for Social Change at the Amsterdam University of Applied Sciences. Until 2021 she held a chair in Media and Cultural Industries at the University of Groningen, and before that worked at Cardiff University and Goldsmiths, University of London. Her work highlights the importance of wonder, doubt, and empathy in understanding current social issues and explores how we can facilitate more inclusive and sustainable societies through creative media and creative research. She is co-author of the book *Beyond Journalism* (with Mark Deuze, 2020).

Marieke Zielhuis

Marieke Zielhuis works at the Research group 'Research Competence' at the University of Applied Sciences Utrecht. She studies the knowledge development and learning processes that take place in collaborations between research, practice, and education. Marieke studied Industrial Design Engineering at Delft University of Technology. Her PhD research focused on the collaboration between design researchers and professional designers and on the contribution of research projects to design practice. She worked in various roles to connect higher education and professional practice: in Centres of Expertise, in education programs for practitioners, and in research collaborations with practice.

Banoyi Zuma

Banoyi Zuma is a changemaker with a focus on participatory design and urban cultural expression. She blends strategic thinking with community-rooted creativity to foster inclusive and impactful change. As Project Lead Samen Dansen at Stichting het Gehandicapte Kind, she heads the national rollout of the inclusive dancing project 'Samen Dansen', a project aimed at connecting children with and without special needs through dance. At Collabros Consultancy, Banoyi serves as Design Lead, crafting processes and narratives where hip-hop culture meets social engagement and institutional transformation. Banoyi holds an MSc in General Management and an MBA in Business Innovation.

405

For Product Safety Concerns and Information please contact our EU
representative GPSR@taylorandfrancis.com
Taylor & Francis Verlag GmbH, Kaufingerstraße 24, 80331 München, Germany

www.ingramcontent.com/pod-product-compliance
Lightning Source LLC
Chambersburg PA
CBHW060751220326
41598CB00022B/2399

9 781041 000204